THE
EXTINCTION
OF EVOLUTION

DAREK ISAACS

BRIDGE
LOGOS
FOUNDATION

Alachua, Florida 32615

So you, son of man, I have made a watchman for the house of Israel. Whenever you hear a word from my mouth, you shall give them warning from me.

—EZEKIEL 33:7

Thank you, Jesus Christ, for your saving grace.

To my wife.
Your example is inspiring.
Your support is unwavering.

My gratitude is also extended to
Christian Lobb, Amy Hammond Hagberg,
and Answers in Genesis.

Bridge-Logos
Alachua, FL 32615 USA

The Extinction of Evolution
by Darek Isaacs

Printed in the United States of America.

Library of Congress Catalog Card Number: 2009911038
International Standard Book Number 978-0-88270-999-4

G163.316.N.m0911.35250

CONTENTS

CHAPTER ONE

THE PRISM OF SIN

Many do not understand the totality of sin and what it did to the world. The original sin recorded in the third chapter of Genesis not only damned mankind beyond any possible human repair; it sent a wrecking ball through our planet and the natural dimension of the entire universe.

The sonic boom of original sin sent a rippling tremor that fractured our world. It created the consummate fight for resources that brings death to the battling animals. It created the struggle that caused some plants to strangle out others for the water and nutrients they need. Sin created the natural war that now engulfs this life. Hurricanes, droughts, earthquakes, and floods are the kicking, violent death throes of the planet. The wages of sin, which is death, is not confined to humans.

That should make complete sense, for Adam was given dominion over all the Earth (Genesis 1:28). Therefore, when his rule brought sin upon the world, all those under his dominion also felt the bitter sting of his wages of sin. The earth is specifically identified as a dying habitat in Psalm 102:25-26:

Of old you laid the foundation of the earth, and the heavens are the work of your hands. They will perish, but you will remain; they will all wear out like a garment.

Therefore, looking with spiritual eyes, we can see that the signs of the Earth failing are the signs of its prophesied death, and we must understand that its fate was sealed the day Adam sinned. We cannot "save the planet" because Earth is in a naturally

1

degenerative state. We can only do our best to act as good stewards before the inevitable end comes.

In saying that, we should be stewards with a clear understanding of our very real circumstances—the saving of mankind is not determined by the health of a planet, which is already lost, but in the spiritual health of our souls, which can be saved.

This thorough understanding of the consequence of the original sin is not commonly taught. What *is* taught about sin is not taught nearly enough, which is the sinful condition of the human heart so perfectly described in Romans 3:9-18. Of that selection, verse 18 becomes most revealing for us where it states of the sinner that *"There is no fear of God before their eyes."*

Understanding that verse, coupled with another key verse in the book of Proverbs will inch us closer to our starting line.

> *The fear of the Lord is the beginning of knowledge; fools despise wisdom and instruction.*
> — PROVERBS 1:7

To many, the word "fear" is taught to be more of a reverence for God than a true fear. I disagree with that assessment. I believe fear in these verses may infer awe, but it also means to be afraid. The words of Jesus reinforce that interpretation.

> *And do not fear those who kill the body but cannot kill the soul. Rather fear him who can destroy both soul and body in hell.*
> —MATTHEW 10:28

There can be no question what kind of "fear" Jesus was speaking of in those verses.

This fear of God, which we know to be the beginning of knowledge, happens when the Holy Spirit presses upon someone and reveals to them that damnation into hell is a certainty for their unrepentant soul. That understanding results in genuine fear. But there is also another kind of fear that many people have experienced and can testify to—the fear of not being God's child, and therefore separated from Him for eternity. These fears trigger a crying out for salvation.

This fear of God, the precursor to salvation, however, is only the *beginning* of knowledge, just as Proverbs 1:7 states above. As Christians mature in their walks, they must move past fear and not allow it to become a long-term layover:

> *For fear has to do with punishment, and whoever fears has not been perfected in love*
>
> — 1 JOHN 4:18B

Understanding our sinful nature and its consequences means recognizing that we are punishable for the sin in our lives. This knowledge of imminent punishment brings fear. However, once we acknowledge that we are rightly punishable for our sinful lives, and ask Jesus Christ to save us from our punishment, we are exempted from that eternal judgment:

> *For God so loved the world, that he gave his only Son, that whoever believes in him should not perish but have eternal life.*
>
> —JOHN 3:16

When a person accepts salvation, there is no longer the fear of punishment. That is why we are taught in 1 John 4:18 that God's perfect love casts out the fear .

The fear of the Lord is the beginning of knowledge, and then at salvation His love perfects us and removes the threat of punishment. We are then freed, mercifully and gracefully from the fear of the judgment, marking the beginning of our journey into wisdom and knowledge.

So why, in this discussion of evolution, is the fear of the Lord so important—so important that we can't possibly go on without understanding it? Why are these Bible verses so critical to understanding this supposedly scientific theory?

The key to that answer is this: if the beginning of knowledge starts with the fear of the Lord, then what does the Bible clearly say about those who have never experienced the fear of God in their lives? It says such a person does not have the spiritual and intellectual platform to begin to gain true knowledge and wisdom.

The most utterly sinful act is to reject the supremacy of God; to reject Him as the Creator of this planet and universe, and consequently, to then deem oneself the highest form of life, intellect, and authority. This condition of the heart is found in the agnostic and the atheist.

The Bible does not mince words when dealing with a person in this state of being. Psalm 14:1 states:

> *The fool says in his heart, "There is no God."*

Those who claim there is no God, or openly admit they have no knowledge of God, can't even begin to comprehend true knowledge and wisdom and therefore can't make sense of the world around them.

That depraved state of mind and soul describes Charles Darwin, the father of modern evolutionary thought, perfectly. He rejected Christ, not on the basis of compelling evidence, but on the basis of Christianity itself when he wrote:

> *I can indeed hardly see how anyone ought to wish Christianity to be true: for if so the plain language of the text seems to show that the men who do not believe, and this would include my Father, Brother and almost all my best friends, will be everlastingly punished. And that is a damnable doctrine.[1]*

Darwin rejected Christianity on the premise of accountability. Note that he did not reject Christ based on scientific reasons; he believed that neither he, nor any of his friends or family, should be accountable to God for any reason.

But calling Christianity a "damnable doctrine" seems quite harsh even for someone in Darwin's frame of mind. What could cause so much animosity? It sounded as if he were motivated to attack God.

The following excerpt was taken from an article pertaining to an exhibit about Darwin's life that was held at the American Museum of Natural History in New York:

> *By the end of the exhibit, however, Darwin is a changed man. He will have found love and married and had children. He will be an agnostic, the last vestiges of his faith stripped away after watching Anne, his 10-year-old daughter and favorite child, suffer from a long-drawn illness and then die.*[2]

I think it is probable that Darwin could not believe a loving, personally involved God would allow his daughter to die. No doubt, Darwin faced a tragedy that most people would not wish upon anyone. Darwin was forced to endure one of the most difficult natural pains known to humanity—the death of a loved one.

However, the death of his daughter in 1851 did not cause Darwin to reach out to God. Instead, he solidified his rebellion against God and the little faith he had was stripped away until he became a full-fledged agnostic.

In 1859, Darwin published his most famous work, *On the Origin of Species by Means of Natural Selection, Or the Preservation of Favored Races in the Struggle for Life.* This book has become the standard for people who want to remove God from the creation account of the universe, and thus remove Him from humanity.

During the majority of Darwin's adult life he treaded a fine, nearly transparent line between agnosticism and atheism. I believe if Darwin were alive today, he would be an atheist; for atheism is the prevailing ideology we now see in those who embrace his theory.

This is important, because Darwin did make vague assertions about a god, or creator. However, due to his agnosticism and obvious distaste for Christianity, those references were made toward a generic god rather than to Jesus Christ. By doing so,

Darwin tipped the initial domino—ultimately leading to the point at which people could attempt to reason God away.

Today, the doors are open in our society to remove God from everything—and core Darwinian principles provide the vehicle to do so. The foundation of atheism that Darwin danced around has become cemented. Evolutionary theory now unapologetically asserts that human life had no creator or designer, and that our lives on Earth serve no designer's purpose—we are simply accidents of nature.

> *Let me summarize my views on what modern evolutionary biology tells us loud and clear...There are no gods, no purposes, no goal-directed forces of any kind... There is no ultimate foundation for ethics, no ultimate meaning to life, and no free will for humans, either.*[3]
> —Dr. William B. Provine,
> Professor of Biological Sciences,
> Cornell University

The fact of the matter is, when Charles Darwin personally rejected Christ, he was forced to view the world through a filter; a prism that separated a personal, intervening God from his view. He consciously made the effort to describe the world, and our humanity, without giving credit to Jesus Christ; for it is through Him all things are made.

I realize that many will argue that it is possible to believe in both positions simultaneously—that man descended from an ape-like creature, as the theory of evolution suggests, and believe in Jesus Christ as God, Creator, and Savior. Many Christians find themselves in that duplicity.

However, as the pages of this book turn, I believe most Christians will abandon that thought and see that those in the middle who try to marry these two foundationally different creation accounts are simply trying to blend oil and water. In doing so, they become the very definition of lukewarm.

Christianity and evolution are mutually exclusive from each other because they originate at radically different starting points.

Christianity starts with the six-day Creation; evolution starts with an accidental occurrence creating a single-celled organism billions of years ago.

With that being the case, it is impossible for the road built from evolution's foundational assumption to ever lead to Jesus Christ as the author of all life. God is removed from the beginning; evolution is not going to graft Him in later.

If someone starts out with a completely faulty foundational assumption and filters all their findings through that faulty assumption, then categorically, everything they discover will be interpreted incorrectly.

Make no mistake—evolution is not just the functional biology scientists learn about when studying living organisms; its principles cannot be used to find cures for diseases and make medicines. Evolutionary biology attempts to figure out how all the species—including humans—came into existence and how they are related to one another. Its core purpose, in fact, is a rejection of God as the Creator:

> *If a moving automobile were an organism, functional biology would explain how it is constructed and operates, while evolutionary biology would reconstruct its origin and history—how it came to be made and its journey thus far.*[4]
> — E. O. WILSON, HARVARD PROFESSOR AND AUTHOR

I find it comical that an evolutionist would state that evolution does not fit into the category of functional biology.

Nevertheless, evolutionary biology reconstructs our origin and history; a purpose that now makes complete sense because Darwin needed an explanation of the world around him once he rejected the Bible. However, that purpose actually leaves evolutionists in quite a dilemma.

Evolutionists believe that all life formed from a single-celled organism. However, everyone knows, including evolutionists, that the theory of evolution is completely inept when it comes to explaining how that "first" single-celled organism came to be and from where the components that supposedly created that life came.

Do not be lured into thinking this answer could be the "Creator" that Darwin invoked sporadically in his writings, because he clearly rejected Christianity. His "god" was merely an idea of some unknown, untouchable force. Furthermore, as we get deeper into the mind of Darwin later on, you'll see that his thoughts were far from what is considered godly.

In that question of origins we find the largest peculiarity about the theory of evolution and its hardened followers. According to E.O. Wilson, a leading evolutionary thinker who has been called a modern-day Darwin, evolutionary biology reconstructs the origin and history of our existence. But, that is the fundamental thing it certainly cannot do.

The origin of our species must start with the origin of that single-celled organism if one is going to believe in evolution, because humans would have eventually descended from that parent. But evolutionists don't know where that organism came from or even what it was. It is just an unknown variable in their concocted equation. Therefore, the very basic premise of evolution is faulty because it fails to explain the most foundational aspect of its theory—the formation of the first living thing.

How can the theory of evolution possibly be trusted to explain how we can grow as a species when it has no idea how we were born? That is like trying to explain how an opera singer can perform an aria without knowing where the sound comes from in the first place!

The theory of evolution is simply flawed—it is that simple. Yet, we find ourselves witnessing a breakdown in logic of shocking proportions. Even though they do not have a clue about where we came from, the evolutionists make bold claims about what we are not:

> *Humanity was thus born of Earth. However elevated in power over the rest of life, however exalted in self-image, we were descended from animals by the same blind*

force that created those animals, and we remain a member species of this planet's biosphere.[5]

— E. O. WILSON

Notice Wilson's foundational belief. He was careful to point out that humanity was born of Earth, not of God; noting that we are a product of the natural, not the supernatural. He rightly followed where Darwin's theory clearly led—to the elimination of God altogether. It is at this point where most evolutionists find themselves today and they cling to evolution as their belief system (and the number of people who embrace pure atheism is only going to grow).

Even though Wilson stated that we were born of Earth, he had no idea how that birth could have happened; for a couple of phrases later he uses a very intriguing term for the source of the beginning of life—"blind force."

Some blind force created the animals, and subsequently created us? It seems the lines have been blurred between science and science fiction, because such a force is not native to Earth, but rather the movie, *Star Wars*!

What would constitute this blind force? How would it be made and what is it made of? Did this force skip over all the other planets in the universe and just create us or did it also create the entire universe? Is it continuing to create planets? Can we see, touch, feel, and pick up this force? And why in the world does it have to be blind?

This blind force is the evolutionist's starting assumption of where we came from. Evolution teaches that mankind descended from an ape-like creature. This ape-like creature would have descended from this undefined single-celled organism. The single-celled organism could not have come into existence without this enigmatic beginning—now scientifically referred to as a blind force. Therefore, defining this blind force would be a critical factor in understanding our origin. Yet, by its nature this blind force, is wholly indefinable!

Let us not forget this single-celled organism that supposedly willed itself into existence at the urging of the blind force—

something its most capable descendent, the human, has no chance of doing. If this was the first living organism, what did it feed upon and where did the food come from? How long did it live before reproducing? How did it reproduce? What was its life span? It certainly would have been difficult to live long and prosper considering its surroundings were completely void of other life. What were the conditions that allowed survival, and how did those conditions arise? Did the organism need oxygen? If so, where did the oxygen come from?

We are constantly told that evolution is based on fact, research, and the evidence all around us. But that is not true. It's based on the belief that a blind force started all life and that the result was our original ancestor. This amazing, single-celled organism that lived in an undefined environment—which is impossible for us to rationalize—eventually reproduced itself, and again we can't be certain how because there is no evidence of any of this.

Talk about blind faith! Yet the evolutionists stand by this completely fabricated theory of origin that is rooted in purely imaginary conjecture. Evolutionists are rolling the dice that some force, which cannot be defined, explained, or even rationalized, made us. They then mock the millions of Christians—who provide testimonies of godly encounters and salvation experiences—for having blind faith. Their hypocrisy knows no bounds.

If this kind of teaching did not ensnare so many people, it would be hysterical. But we can't laugh at it, because despite the absence of true wisdom, which stems from the lack of fear of the Lord, evolution is the most threatening social movement in the world today.

As I alluded to previously, I am well aware that some would stubbornly try to pose the argument that God was the "blind force" in evolution and it was God who created that first single-celled organism billions of years ago, which then triggered the evolution of the species from that point on.

Adding a human interpretation of a creator to the evolutionary equation does not work out. I say "human" interpretation of God because we would have to reject the Bible's account of Creation to accept the evolutionary story. Therefore, we would make God in our image, He would have to operate under our guidelines, and He would have to create the world, and consequently us, in a way that meets our approval.

In other words, in this God-based evolution scenario, the Creation is telling the Creator, how He made us. This reduces the vastness of God, the method of God, and the way God intervenes in humanity to whatever their finite, limited minds can decipher. In doing so, proponents of evolution promote themselves as the authorities on how the Earth was created even though they have no ability to create an Earth for themselves.

But there is more to it than that. To add the designer component to the theory of evolution would mean to also add the designer's purpose to evolution. Anything that is specifically designed is designed to accomplish a planned purpose and to fulfill a very specific desire of the designer.

Adding a designer's purpose, however, puts the *kibosh* on the core evolutionary principle that the development of species is driven by accidental, random mutations sifted out by natural selection. There is nothing accidental about a design, or the results of a design. A design is created to produce an intended outcome. A design is planned and orchestrated with attention to detail.

Therefore, the evolutionary idea that all of nature, including humans, is a product of unplanned accidental events would be a contradiction to the pointed, exact outcome of a designed existence.

Adding God as the designer causes irrevocable conflict with the accidental ways of evolution. That is why it has become critical to evolutionists to insist on a blind force, and not a designer. E.O. Wilson was actually fundamentally correct in claiming a blind force created us—if one is to believe we are products of evolution. The theory of evolution demands that life was birthed from an undefined, undiscovered, accidental, and imagined starting point. There is no other possible foundation that evolution can be traced back to, and due to that, we quite easily see the frailty of the theory. It stands on a mere whim of imagination.

The truth is Darwin rejected God. When one has no knowledge of God, when one is still bound and guilty to the sin of Adam, when a heart rejects the foundational truth of the Creator God, they can glean little wisdom from this world that is persistently steadfast and true. Such a person, as Darwin was and as atheistic evolutionists are today, sees the entire world through eyes that have been compromised by sin. They see the world through a Prism of Sin.

A prism is a glass or other transparent object that separates and refracts white light into multiple colors. A Prism of Sin is similar. Even though evolutionists see the same world Christians do, their view becomes fractured, distorted, and confused because an inherent sin nature fractures all that they see. By rejecting God they end up believing in a blind force that created a single-celled organism. There is not a shred of proof for either one. Sin deceives and gives frailty the illusion of strength.

Evolutionists then try to interpret what sin has already distorted while continuing to reject the authority of God and His word. No truth can come from this sinful condition—only a severe distortion of the truth.

This distortion, which manufactured a blind force origin, coupled with the recklessly speculative first single-celled organism, is a reason why evolution has been so hard to argue from a scientific perspective. Science is supposed to be based on research, findings, and fact. These pillars of science however, are absent on the blind force origin and the first single-celled organism. Therefore, evolutionary theory cannot function as a true scientific cornerstone because it is as stable and dependable as a log adrift in a river.

For this reason, scientists who attack evolution with actual science are trying to hit a moving target. The science of evolution is really only smoke and mirrors; the precedent has already been set that mere speculation, imagined origins, invented environments, and the written story lines of supposed lifestyle habits can be passed off as scientific evidence, where no proof is necessary.

Imagination, to some degree, is essential for grasping key events in cellular history.[6]

—B.D. DYER AND R.A. OBAR

Therefore, to effectively discuss evolution and what it means to humanity, we need to look beyond evolutionary "science." We need to look at the social aspects of evolution and its spiritual ramifications.

A social approach would ask us if evolution is true, then how does it impact our lives? To discover the answer, we cannot categorically reject evolution, but must actually embrace it to a point. We will let evolutionary thought and doctrine take over the reins, and we will sit back and let this horse run its course. That is the premise of *The Extinction of Evolution*.

Following our social dissection of evolution, we will look at it through spiritual eyes. After we see the results of the social outcome, the spiritual reality of evolution becomes very clear. It is not what it claims to be—the reconstruction of our origins—but it is something else. Our spiritual exploration into evolution will reveal exactly what it is, why it is here, and where it is going.

To aid in these approaches, I created a character named Dr. Iman Oxidant. Dr. Oxidant is a full-fledged, atheistic evolutionist. He believes all rational people of the world understand that evolution is a fact, and that it is as certain as gravity.

Through his lectures, Dr. Oxidant pushes the world to fully embrace evolution and take it out of the textbooks and into the streets. He may be the most honest evolutionist you'll ever meet; he follows the ideology of evolution to the bitter end.

Most things born from sin seem quite harmless, reasonable, and even alluring at first. You might find Dr. Iman Oxidant to be the same. But things born from sin do not just disappear or leave well enough alone when a simple deception has occurred. Sin brings death and destruction. A worldview built upon a foundation of sin will ultimately bring great evils upon the world. As Christians, we must understand this possible outcome.

ENDNOTES

1   Darwin, Charles, *From So Simple a Beginning*, New York: W.W. Norton & Company, Inc., 2006, pg.1482.

2   Than, Ker, "The Life of Charles Darwin: From Aimless Adventure to Tragedy and Discovery," LiveScience, November 16, 2005.

3   Patterson, Roger, *Evolution Exposed*, pg. 82.

4   Darwin, Charles, *From So Simple a Beginning*, New York: W.W. Norton & Company, Inc., 2006, pg. 12.

5   Ibid., pg. 12.

6   Patterson, Roge, *Evolution Exposed*, Pg. 149.

CHAPTER TWO

THE BIOGRAPHY OF
DR. IMAN OXIDANT

Iman Sliyt Oxidant was born on October 12, 1941, to Gert and Gertrude Oxidant, in Vereenigeng, South Africa. The Afrikaans family moved to London, England when Iman was twenty-one years of age. An accomplished student, Iman studied at the finest academic and research institutes in Europe and earned advanced degrees in theology and philosophy. He also studied sociobiology, but was never awarded a degree in that field. Oxidant's degree in theology is similar to the one received by Charles Darwin from Cambridge University.

In 1996, Dr. Iman Oxidant moved to Boston, Massachusetts, and founded The Institute of Progressive Lineage. The Institute's mission is to promote the evolutionary lifestyle to all of humanity. This includes teaching the masses to disregard dated and dangerous religious rhetoric, as well as adopting scientific reasoning and data as the standard for all societal foundations.

Dr. Iman Oxidant was commissioned by The Institute of Progressive Lineage to conduct a series of web seminars promoting the Sub-Laws of Evolution. The Sub-Laws of Evolution are four doctrinal truths that humanity must live by and understand if evolution is to be regarded as law and therefore applied to our lives.

While in Boston, Dr. Oxidant broadcasted eleven online seminars on the four Sub-Laws of Evolution from his wine cellar. The seminars have become historically significant and are considered scientific canon in today's evolution-driven world. Consequently, word-for-word transcripts of the seminars are available, without any editing or corrections, for your review.

Ironically, Dr. Iman Oxidant was never able to see the impact of his work. On November 24, 2007, at the age of sixty-six, Iman tragically lost his life at the hands of his youngest son. According to the police report, Ivan Oxidant, upset that he had been left out of his father's will, killed his father and two older brothers while they were sleeping. After the disappearance of Dr. Iman Oxidant's wife, Stella, on December 25, 2007, Ivan became the sole heir to his father's substantial fortune.

Despite his premature death, Dr. Iman Oxidant's examination of the Darwinian lifestyle lives on. Here at the Institute we are proud to preserve Dr. Iman Oxidant's legacy and present his work to you in its original form. We are certain his words will help you find inspiration in this completely accidental thing we call life.

Warm regards,

Leonard Huxley, Esq.
Interim Chairman
The Institute of Progressive Lineage

CHAPTER THREE

FOUNDING BELIEFS

From the cellar of Dr. Iman Oxidant

GREETINGS, people of science and all rational minds alike:

I am your instructor and mentor, Dr. Iman Oxidant, the founding member of The Institute of Progressive Lineage. The Institute is a scientific think tank devoted to the implementation of science into everyday life. We "socialize" science.

At the Institute, we have decided to take our fortuitous existence seriously. We have jerked our fate away from the imaginary hands of the gods, creators, and cosmic designers; we believe humanity is finally ready to feel the pride of simply being animals—and nothing more.

I am assuming because you are committing yourself to be under my intellectual tutelage, that you understand a basic truth, that all things of value can be discovered, derived, and found through science.

Science reveals the spectacular and beautiful randomness of nature. Science dispels the myths and measures the validity of truth. Through science, we discover the mysteries of our planet and the magnitude of the accident that created the human species.

Every product or practice found its origin in the mind, hands, and actions of a scientist. The chair you sit in, the sidewalk you walk on, and the car you drive, all have origins in science. If the physical sciences have an impact on our daily lives, evolution, the science of our origin, must impact us even more. Evolution gives us our creation story. Evolution explains that mankind was not independently created in the image of some mythical god, but rather that we are descendents of some ape-like creature. We remain a mere extension of the evolution of unintentional animal life.

Because evolution helps us understand our past, it can help us plan for the future. But before we can scale a mountain, I need to make sure all of you have the ability to put one foot in front of the other. Therefore, since I am assuming most of you have not had a biology class in the last few days, we need to explain evolutionary biology; for the definition of evolution also evolves. It is an ongoing joke that if you put ten evolutionists in a room, you will get eleven different definitions of evolution. In light of that, we have done our best to sum it up. Accept my apologies, my evolutionary brethren, if we left something out.

Evolution through natural selection was a theory presented by Charles Darwin that asserts heritable changes in a population, through pressures of natural selection lead to the physical change of skeletal and molecular structure in an individual.

Evolution explains the great diversity of life on this planet. This theory allows us to believe that non-living chemicals, of origins unknown, at some point in our past, exploded into life. This birth of life is scientifically explained as a blind force that created the infamous single-celled organism, which spontaneously, unintentionally, and remarkably jumpstarted life on this planet.

Eventually, after billions of years of natural selection, complex life was formed. It was from these humble beginnings that humans eventually entered the picture.

While the concept of evolution floated around the scientific community for some time, it was not until English naturalist Charles Darwin unveiled his theory of natural selection that it really took hold.

The term "natural selection" is used almost interchangeably with the phrase "survival of the fittest." These two terms have come to represent what the scientific community understands to be the mechanism of evolution. It is true that random mutation aids evolution, but it is natural selection that chooses which mutations are fit to survive.

Natural selection favors individuals who are better adapted to their environment. In contrast, it is relentless in its gradual elimination of those individuals who are less adapted and less suited for survival. It is important to understand that even though the winnowing process of natural selection is ever-present and the weak are constantly being siphoned out, there will never cease to be those who are weak in comparison to the strongest of their species. Each generation must contain a group of unfit individuals in order for natural selection to continue operating.

This brings us to the obvious conclusion that we must, as honest evolutionists, openly embrace. Humanity is a product of evolution. Therefore, there must be populations within the human species that are unfit for survival or natural selection is powerless and can do nothing. Natural selection favors the strong, but it does so at the expense of the weak.

Our challenge as evolutionists is to take this knowledge, and the scientific idea of evolution, and provide real-life applications of its meaning to humans. This is long overdue. Unfortunately, this kind of real-world application, where one can see the true effects of evolution, rarely happens. Most educational institutions focus only on a small fraction of the whole—a mere sliver of the comprehensive theory of evolution. This is a terrible disadvantage to the students.

Today's youth are not being given the opportunity to see the full capability and true nature of evolution, and consequently, the true nature of their bodies and instincts. These students profess to believe in evolution, but do not truly understand what they are professing. This is a state of mind we will attempt to remedy.

One of the keys to accomplishing this is to go straight to the source of evolutionary thought. We cannot expect our younger generations to properly understand evolution if they do not understand the foundation that it has been built upon. That foundation is of course, the work, research, and writings of Charles Darwin.

Much of our research comes from Darwin's book, *On the Origin of Species*. E.O. Wilson, a professor of evolutionary biology at Harvard University, once said that this work of Darwin's is

"… arguably history's most influential book"[1] and what "… many scholars consider the greatest scientific book of all time."[2]

It is interesting to note, however, that the title of this groundbreaking book was shortened. The original title brings a quicker enlightenment to the subject matter than the amputated version we now use today. *On the Origin of Species* was first published in 1859 under the title, *On the Origin of Species by Means of Natural Selection, or the Preservation of Favored Races in the Struggle for Life.*

Preservation of Favored Races—The first brick of the evolutionary foundation has just been laid, my young pupils. It was this work, and others, that E.O. Wilson, claimed "spread light on the living world and the human condition."[3]

I trust you are starting to have a clearer picture of why there will always be individuals within a species, who by nature are unfit to survive. They are unfit, or *non-favored*, simply because of their genetic identity, and consequently, their physical and intellectual make-up.

Wilson further elaborated on Darwin's scientific reach when he said:

> For over 150 years his books…have spread light on the living world and the human condition. They have not lost their freshness: more than any other work in history's scientific canon, they are both timeless and persistently inspirational.[4]

If something is timeless, that means it is as true today as the day it was forged. Therefore, sinking our intellectual teeth into the mind of Charles Darwin will reap huge benefits toward helping us see how humans fit into the natural world. Knowing what Darwin believed will help us understand why and how he discovered evolution through natural selection, and through that, deepen our own understanding of our origins and enable us to plan for the future.

This almost goes without saying, but those of us who understand Darwinian thought have an enormous burden upon our shoulders, for we should be the ones who set the course for the future. It only makes sense to build our society from accurate evolutionary blueprints. Those of us embracing science and reason should be the executives who envision our future. If we do not rise to that challenge, we will be responsible for our own failure—a risk I am not willing to take.

The opposite of those on the forefront of science are the vast seas of the irrational religious sects. Everyone must realize that the inaccurate understanding of our origins by the religious masses will lead to catastrophic outcomes. In a world where man now has the ability to destroy the Earth in numerous ways, it is much too dangerous for the human species to continue to be guided by misnomers, myth, and unsubstantiated legend. We cannot allow the fate of all mankind to be held in the hands of the primitively ignorant.

As the dominant species on Earth, our fate rests squarely on our human ability to deduce and reason; to contemplate, analyze, plan, and then act. We believe our fate is in our own hands and not in the hands of some imaginary being that we can neither, see, touch, hear, nor feel. Humans are the most evolved species on this planet; there is no higher intelligence governing us. We must rely on the reason and intellect we can squeeze out from our own collected thoughts. We must embrace our dominance and refuse to bow down to the manufactured morality of those of our species who are clinging to the past.

Failing to act and failing to live according to the knowledge gathered by evolutionary science will lead us to the guillotine of extinction. In the end, we will share the fate of the dinosaur and all the other dominant, yet extinct species before us. As responsible, evolutionary-driven individuals, we cannot allow the ship to sink because the monkeys of our species continue to drill holes in the hull.

My fellow humans, we are at a crossroads in our history. Evolution has given mankind a choice; we can continue to waft in ignorance, or we can prove we are strong enough to survive. We can allow astute scientific minds to lead us into the next stages of evolution, or we can allow the ignorant, those steeped in archaic beliefs, to lead us toward extinction.

The answer is obvious. We must use evolutionary science as the platform for our lives, communities, and the entire global civilization. This our Founding Belief.

ENDNOTES

1 Darwin, Charles, *From So Simple a Beginning*, New York: W.W. Norton & Company, Inc., 2006, pg.11.

2 Ibid., pg. 437.

3 Ibid., pg.11.

4 Ibid., pg.11.

CHAPTER FOUR

THE SUB-LAWS OF EVOLUTION

From the Cellar of Dr. Iman Oxidant

> *It is a truly wonderful fact ... that all animals and all plants throughout all time and space should be related to each other...*[1]
>
> — CHARLES DARWIN

MY EAGER MINDS and itching ears:

I would like to briefly discuss this quote from Charles Darwin, if I may. As is my style, we will dissect his words so that no mistake can be made in the conclusions we draw.

If all organisms are related, that means every living organism, be it mammal, plant, insect, or fish, stems from the same parent; conceivably billions of years ago. How intriguing it is that we are kin to both the horsefly, and the horse it bites.

Random chance orchestrated this development through the workings of evolution. Random events, guided by nothing, with no goal or purpose, formed life over a great chasm of time. This sequence of random, accidental events has become our real creation story for all life on the planet, including the human species. This creation story is far superior to the myths found in ancient cultures; especially, the Bible.

Humans are classified as *Homo sapiens sapiens*, a species in the animal kingdom. Therefore, using very simple logic, humans are animals and, are therefore, related to all other animals and all plants throughout the history of this planet. Humans are related to monkeys, whales, and cacti; we are not, as religious zealots would believe, made in the image of god.

I want to be very clear about that. No matter how elevated we seem to appear, we are still animals that came from the same organic matter that created every other living thing on this planet. In a very real sense, Mother Nature is mother to one family, one very large family. Humans are merely the strongest "siblings" of the litter—for now.

In his work, Charles Darwin identified this family and illustrated how all species branched out from one common ancestor, or trunk. He named this analogy the "Tree of Life," which connects all living creatures together. (By the way, this is not the same "Tree of Life" mentioned in the Genesis story found in the Bible. We can safely assume, however, that Darwin, who had a degree in theology, was mocking that myth, for there are many other names he could have chosen.)

Over the course of history, numerous species have died out and faced extinction. Others, however, have prospered in the vigorous competition ever-present in nature, and continue to sprout and create diverse forms of life. As the tree grows and produces innumerable branches, each branch represents a new species; a species that diverges in character (a change in their physical makeup) away from their parent. It is actually this divergence of character that allowed an ape-like creature to gradually become man.

These concepts, which help describe evolution through natural selection, have long been thought of as the theory of evolution. But because of the strength of Darwin's reasoning, evolution is now recognized as much more than just a theory. According to E.O. Wilson:

> *Evolution by natural selection is perhaps the only one true law unique to biological systems ... and in recent decades it has taken on the solidity of a mathematical theorem.*[2]

Evolution has been so well described and so solidly established that we can now consider it a fact of nature. Evolution operates through the laws of nature with great certainty; it is as certain as gravity. This should not come as a surprise to anyone. Our schools, universities, and the educated population all teach and profess evolution to be fact. Its support in academic circles is nearly unwavering.

Nevertheless, the stunted intellects of those who oppose scientific progress are becoming a threat to our advancement as a species and a culture. We must deal with the errors of the masses. We have allowed the uneducated to go on witch hunts, sail all over the world and call it flat, and continue to think the sun revolves around the Earth. We can no longer do that. We must deal firmly with the irrational minds that oppose us. Too long have we allowed outdated mythologies, soothsayers, dead prophets, and hypothetical morals derived from religion to impact our modern society.

I believe it is time for principles, derived from evolutionary science, to become the guiding light of humanity. The world is becoming a practical place, and practical ideologies need to rule the day.

Evolution through natural selection is the story of humanity. It alone is the active mechanism that has dictated the formation of our lives. Understanding how it works, and how our lives are impacted by it, can allow the most advanced humans to stay one step ahead of the swinging sword of natural selection and continue to be members of the dominant species on the planet.

One thing humans must understand in this pursuit is that evolution through natural selection is a non-forgiving natural government. It rules with absoluteness and there is no leniency or mercy. We can see that by taking a cursory look at our extended family. If two lions are living in a habitat in the plains of Africa that is devastated by drought and come upon a dying gazelle, there will be a fierce competition for the right to feed upon that gazelle. For both lions, the need to eat the gazelle is dire. We could say the environment is so extreme that the lion that eats the gazelle will survive and the one that does not will die from weakness and starvation.

At this point, natural selection is in full swing; it will choose which lion is the strongest and which is the weakest, which has the right to continue and produce more offspring and which will perish. Natural selection will decide which genetic line will continue to propagate itself and which will not. This choice is manifested through the actions and reactions of the lions. Their actions are enabled by their unique genetic identity. The lions will fight each other for the right to eat the dying gazelle. The victor of this competition will be chosen by natural selection for survival.

Like the lion, it is our reaction to the environment that determines how, and if, we survive. Our actions, enabled by our genetic identity, will determine our personal resources, lifestyles, and most importantly, reproduction habits.

Our physical and intellectual reactions to weather, disease, food supply, and aggression from other animals demonstrate whether we are of strong or weak genetic makeup. If one is strong, then one survives. If one is weak, then natural selection exposes that weakness in key survival situations—breeding, eating, and euthanizing—and ultimately eliminates the weak from the gene pool of the Earth. This will make the human species stronger.

This government of nature, though it can seem complex at times, is very simple in its core. Darwin explained natural selection succinctly and perfectly when he summarized, "let the strongest live and the weakest die."[3]

Because humans are products of this natural world, the laws that govern it—primarily natural selection—exercise an absolute authority over us whether or not we want them to. The natural laws that have shaped the very fingers at the tips of our hands have the kind of authority we cannot ignore. We are completely enslaved to it and formed because of it.

On this cause, we believe we can adapt strategies and directions based on our knowledge of evolution that can guide us into greater prosperity as a species. Part of doing this may be exposing fallacies of other competing thoughts and ideologies; for we cannot allow outdated and clearly errant thought to risk the dominance of humans on this planet.

My eager students, since we have now evolved to the intellectual level to accumulate all of this knowledge, it would be absurd to try to live our lives as if evolution were relevant only in textbooks. We must live out natural selection and survival of the fittest as if they were law. Evolution has governed the natural world since the original ancestor gave birth to all life on this planet, and it still dictates the natural progression of life. Understanding its ways can help us secure a strong and safe future for our species and for our accidentally formed planet.

Herein, we reach our dilemma, my scientific peers and underlings. At present, evolution does not rule our streets, natural selection is not occurring in our homes, and survival of the fittest is seen only on the nature channels.

A cursory look at modern humans shows that the mainstays of evolution are absent in our human habits. If we don't show the mechanisms needed for evolution, then how do we evolve? That is an astute point, and we can only conclude that this situation will only be allowed to continue for so long. Nature will move against us for our slothful excess, and we stand in mortal danger.

Natural selection has taken billions of years to carefully craft the human mind and body into the dominant life form. It stands ready to move against us if our loathsome inefficiencies remain unchecked and unchallenged. If you believe in evolution, then you join the entire scientific community in understanding that I am correct. The writing is clearly on the wall and only the religious zealots oppose it.

We have reached a point in human history when a terrible sword is coming down upon us. Do we live in the past, driven by religion, mythology, and fallacy? Do we revel in all the inefficiencies and excesses that come with worshiping an imaginary god? Or do we turn to science as the guide to lead us forward?

The answer is obvious—science. The science we turn to is that of our origin—evolution. For it is only through proper understanding of our past that we can navigate the present and plan for the future. We must become intentional about an evolutionary future and advance our species to its pinnacle.

To accomplish this task, and to give our species hope, we at the Institute have created the four Sub-Laws of Evolution. These Sub-Laws are the lines of reason and logic that are drawn directly from the kernels of Darwinian thought. To create these Sub-Laws we stayed truer to Darwin than Christians stay to their Bible. By using Darwin as the foundation of all our thought, and adding to that the superior understanding of modern evolutionary science, we stay above reproof. We are here to end debate, not create it.

These Sub-Laws are the common sense outcomes of Darwin's evolutionary legacy and should lead to great prosperity when humans, as a species, walk upright and in perfect harmony with the principles of natural selection. They provide a template for human action and behavior and outline a definitive standard based on science and progress—not on legend and misconception.

Without further ado, follow me into the real world of evolution. Let us observe it not from the acute angle of mere information gathering, but from the 360-degree, all-encompassing, bird's-eye view of the world according to Darwin. Let the games begin, and let the best man win.

ENDNOTES

1   Darwin, Charles, *From So Simple a Beginning*, New York: W.W. Norton & Company, Inc., 2006, pg. 532.

2   Ibid., pg.11.

3   Ibid., pg. 605.

CHAPTER FIVE

THE SUB-LAW OF THE MYTH OF HUMAN EQUALITY

From the Cellar of Dr. Iman Oxidant

MY DEAR LUMPS of organic matter:

Many of us possess unquestionable genius. Professors, scientists, and think tank minds have reached great heights.

Our strength, however, is not shared by the entire species. The human species is partitioned between the intellectually strong and the intellectually weak, and between those with superior genetics and those with inferior genetics. Unfortunately, the weaker strains continue to impede our progress.

Oftentimes, those who lag behind hold onto mythological belief systems. Their intellectual evolution allows only the comprehension of errant religion and does not give way to the heightened aptitudes needed to understand science and to grasp deep thought. Certainly, they are nowhere near the depth to which my mind can descend.

These weaker strains of humanity have made it through the sieve of natural selection because of our behavioral error of not following our evolutionary instincts. Their very existence is a weakness plaguing our species that will leave us vulnerable to species-wide action by natural selection if we allow them to continue existing. This segregation in our species is unmistakable, and through the very words of Charles Darwin I will prove to you, beyond a shadow of a doubt, that all humans are not the same. We are partitioned—equality is a myth.

Darwin clearly defined who the more capable humans are based on gender and race. He first alluded to this certainty when he used the phrase "favored races" in the title of his heralded book, *On the Origin of Species by Means of Natural Selection, or the Preservation of Favored Races in the Struggle for Life.*

His use of this explicit language was intentional. Due to his unapologetic clarity, I can now begin to present to you what evolution really means with the Sub-Law of the Myth of Human Equality.

Rational people would contend that the idea of human equality must somehow stem from religious beliefs. I am drawing specifically upon Christian doctrine to form this conclusion. I can't argue with the fact that compassion, coupled with a belief in equality for all, exists in many cultures, so it must have come from somewhere. If these human tendencies do not stem from Christianity, then they have to stem from some other religion, philosophy, or non-Darwinian belief.

In recent history, the notion of equality was finally cemented into modern culture by governments that stated that all men are created equal. So religion, philosophy, or myth—whichever term may be applied— has left us in the rut of ignorance from which we are now forced to extricate ourselves.

To start the proverbial purging, let me say that the idea of *equality* is not a scientific position. The concept of equality, as I will make painfully clear, is not only absent in evolutionary thought, it fundamentally contradicts it.

Our understanding of Darwin's evolution through natural selection has led us to conclude that "some organisms in a species have certain inherited variations that give them an advantage over others."[1]

Because the Sub-Laws are about human survival and human evolution, we will be applying Darwin's evolutionary law solely to the human species; an approach which is long overdue. Therefore, let us insert the word "human" in the quote for the sake of clarity:

> *Some organisms in the human species have certain inherited variations that give them an advantage over other humans.*

This statement confirms that some individuals within the species of the human race have attributes that give them advantages over others. Simply put, humans are *not* equal; over 6.5 billion humans cannot be evolved at the exact same level—it is statistically impossible. These advantages are what the process of natural selection uses to decide between one individual and another while they struggle for existence.

This struggle for existence is critical with the constant influx of people throughout the planet. It must act as a siphoning process that destroys the weak and thereby protects the planet from overpopulation and the stripping of its resources.

> *Hence, as more individuals are produced than can possibly survive, there must in every case be a struggle for existence, either one individual with another of the same species, or with the individuals of distinct species, or with the physical conditions of life.*[2]
>
> — CHARLES DARWIN

This struggle inevitably produces a victor and a loser. Less improved individuals of any society are the ones sifted out of the population during this struggle. Darwin went beyond theory and noted the groups of people he believed were among the lowest evolved on the planet when he wrote:

> *I believe, in this extreme part of South America, man exists in a lower state of improvement than in any other part of the world.*[3]

During his travels on the *HMS Beagle*, Darwin was able to observe many races of people. Through his unique scientific method, he was able to conclude that the "man" that inhabited a particular region of South America—the one he called the

"Fuegian," was the least improved of all men on the planet. Being the father of evolutionary thought, Darwin clearly meant that, from an evolutionary standpoint, the Fuegians were the least genetically advanced race.

Fortunately for us today, the genius of Darwin did not stop there. He continued to list the other races of man that were in some state of lower improvement than the race of Darwin:

Some of the tribes of Southern Africa, prowling about in search of roots, and living concealed on the wild and arid plains, are sufficiently wretched.[4]

Charles Darwin once again gave us good food for thought. Speaking of the indigenous tribes of southern Africa, he described those people as "sufficiently wretched."

Darwin described the eating and living habits of the Africans to be like those of animals when he said, "prowling about in search of roots." These are hardly the words of a man who thinks he has met his evolutionary equal. They are often called "black people" today, but Darwin used a special name for them throughout his writings—"Kafirs."[5]

Very few people outside the continent of Africa understand what that word means, so allow me to enlighten you. The root word of Kafir, spelled both Kafir and Cafir, comes from a Muslim dialect and means "infidel." When the Europeans colonized South Africa in the nineteenth and twentieth centuries and instituted apartheid to keep the black tribal people impoverished and used for the equivalent of slave labor, the Dutch Afrikaners called the black Africans, "Kafirs." This word is regarded as highly offensive today and is actionable in South African court if used. One modern meaning of the word is a person without a soul.

Even today the term creates controversy. In the 2006 Hollywood film *Blood Diamond*, the main character shouts out the word "Kafir" to his black companion and a fight immediately follows.

Darwin continued to categorize the races of men:

The Australian [Aborigine], *in the simplicity of the arts of life, comes nearest the Fuegian: he can however, boast of his boomerang, his spear and throwing-stick, his method of climbing trees, of tracking animals, and of hunting. Although the Australian may be superior in acquirements, it by no means follows that he is likewise superior in mental capacity; indeed, from what I saw of the Fuegians when on board, and from what I have read of the Australians, I should think the case was exactly the reverse.*[6]

The South Sea Islanders, of the two races inhabiting the Pacific, are comparatively civilized. The Esquimau in his subterranean hut, enjoys some of the comforts of life, and in his canoe, when fully equipped, manifests much skill.[7]

Darwin showed us that these different races of people exhibited different levels of improvement. However, in these lesser-improved forms of life, a consistent trait is exhibited. As anyone can see, all these groups have higher levels of melanin in their skin than what you see in Darwin's white European race. Darwin was so certain of the differences in men, and so certain these were lower forms of life, that he provided us with other insightful looks at his research and travels:

While going one day on shore near Wollaston Island, we pulled alongside a canoe with six Fuegians. These were the most abject and miserable creatures I anywhere beheld.[8]

Here, we see Darwin called them creatures, and miserable creatures at that.

Viewing such men, one can hardly make one's self believe that they are fellow-creatures, and inhabitants of the same world.[9]

Darwin understood that what we traditionally have thought of as fellow humans may not be human at all. Let's continue and further develop this line of thought:

> *Their skill in some respects may be compared to the instinct of animals; for it is not improved by experience: the canoe, their most ingenious work, poor as it is, has remained the same, as we know from Drake, for the last two hundred and fifty years.*[10]

As we are able to see, these tribal people had skills that were similar to the instinct of animals. Darwin's quote begs the comparison of these canoe-building people to the dam-building beaver. A beaver can build a dam but it does not have the ability to ever improve on that design. The technology of the dam is limited to the instinct of the animal. Much like beavers and other animals that build simple structures, these dark-skinned tribal people could build a canoe but did not have the ability to improve it.

At this point, Darwin is starting to formulate and solidify his evolutionary theory and understand that there may indeed be *favored* races. However, that may be rash to claim just yet, so let's dig into more of Darwin's mind:

> *One of our arms being bared, they expressed the liveliest surprise and admiration at its whiteness, just in the same way in which I have seen the ourangoutang do at the Zoological Gardens.*[11]

Once again, Darwin delivers his thoughts with such relevant language! He noticed that these less-improved forms of man acted much like primates at the zoo. Just like primates, these people were very impressed by the whiteness of his skin; almost as if they wished their skin was like his. To be honest, I would be fascinated to know how Darwin knew what the jumping orangutans at the zoo thought of his skin; it would be interesting to study his research on that.

As we continue, Charles, boasting a most brazen form, brings a masterfully loaded question:

> *It might also naturally be enquired whether man, like so many other animals, has given rise to varieties and sub-races, differing but slightly from each other, or to races differing so much that they must be classed as doubtful species?*[12]

Darwin cleverly seeds the thought that some races differ so much that it is doubtful they are members of the modern human species at all. Based on Darwin's thoughts about the differences between the races, one might even conclude that interracial marriages should instead be called "interspecies" marriages.

Darwin has asserted what evolution demands—that some humans may only be subhuman, while others may not even be human at all. I would attest that through his powerful use of language, Darwin believed that the Fuegians, South Africans, and Aborigines of Australia might not even be in the same species as European humans.

Because Darwin has helped us understand the differences in the races we owe him much. And after all these years we can rest assured that his train of thought is not abandoned; it is still relevant and used as the foundation of the biological sciences:

> *For over 150 years his books … have spread light on the living world and the human condition. They have not lost their freshness: more than any other work in history's scientific canon, they are both timeless and persistently inspirational.*[13]
>
> — E.O. WILSON

Accordin to Harvard professor, E.O. Wilson, these words of Darwin still apply today—and he finds them persistently inspirational. I don't know about you, but as a dominant white male of European descent, I feel inspired.

Darwin continued to make distinctions between the highest and lowest races of people; those who are human and those who only masquerade as human:

> Nor is the difference slight in moral disposition between a barbarian, such as the man described by the old navigator Byron, who dashed his child on the rocks for dropping a basket of sea-urchins, and a Howard or Clarkson; and in intellect, between a savage who does not use any abstract terms, and a Newton or Shakespeare. Differences of this kind between the highest of men of the highest races and the lowest savages, are connected by the finest gradations.[14]

Darwin clearly places the Caucasian male at the top of the evolutionary food chain. He does this by including men, like Newton and Shakespeare in his comparison between the other races of men he simply calls the "lowest savages." Do I think it was a coincidence that Charles Darwin was of the same race as Newton and Shakespeare, what he called the highest of races? Yes I do. Darwin was a scientist and scientists never allow personal motivations or opinions to influence their research.

I do understand that many of you are not of the highest race and this evolutionary truth might cause some growing pains in the world of the subhuman. However, you can't fight science. Please don't get angry with the messenger; I am only relaying the evolutionary truth about your origin, and in time you too will find it inspirational.

And one more thing, if you are of the lowest race, well then, maybe you should start going to the veterinarian instead of the M.D.

I think we have done enough damage here. Let's continue in a slightly different direction.

Darwin explained that "all organic beings are exposed to severe competition."[15]

Humans are residents of this planet. Our actions within this competitive environment are part of the winnowing process of natural selection—the same way a lion kills another lion to gain control of the pride, a pack of hyenas chases off a single cheetah, a dog kills a cat simply because of an instinct to kill, or a man pursues a woman for mating purposes. All of these are actions of animals in nature—the actions of natural selection and survival of the fittest. The lion, hyena, cheetah, and human do not know they are carrying out the designs of natural selection when they act upon their instincts and fulfill their basic urges.

Nevertheless, those instinct-driven actions illustrate the power of natural selection. Nature is not kind, nor is it peaceful; the survival of some depends on the destruction of others. If the lion ever stopped fighting for his pride, he would be "naturally selected" for death by a more vigorous male that would kill the more docile and lazy lion and take over the pride. Darwin explained this evolutionary principle best:

> *Nothing is easier than to admit in words the truth of the universal struggle for life, or more difficult, at least I have found it so, than constantly to bear this conclusion in mind. Yet unless it be thoroughly engrained in the mind, I am convinced that the whole economy of nature, with every fact on distribution, rarity, abundance, extinction, and variation, will be dimly seen or quite misunderstood. We behold the face of nature bright with gladness, we often see superabundance of food; we do not see, or we forget, that the birds which are idly singing round us mostly live on insects or seeds, and are thus constantly destroying life; or we forget how largely these songsters, or their eggs, or their nestlings are destroyed by birds and beasts of prey; we do not always bear in mind, that though food may be now superabundant, it is not so at all seasons of each recurring year.*[16]

Darwin's point here is very clear: the physical world we live in is not a fairy tale or cartoon. Nature is a harsh environment where the destruction of some organisms benefits the continuance of others. Nature is about life and death, strong and weak. It is this cycle that makes evolution possible.

One of the core elements of evolution is that the fundamental competition for survival is most intense within the same species—just as a stronger lion kills a weaker one to gain control of the pride. Nature not only allows—but actually promotes—such killing. In the human species, the competition for survival is likewise most severe with other humans. Darwin states:

> [A]s *new species in the course of time are formed through natural selection, others will become rarer and rarer, and finally extinct. The forms which stand in closest competition with those undergoing modification and improvement, will naturally suffer most. And we have seen in the chapter on the Struggle for Existence that it is the most closely-allied forms, varieties of the same species, and species of the same genus or of related genera, which, from having nearly the same structure, constitution, and habits, generally come into the severest competition with each other. Consequently, each new variety or species, during the progress of its formation, will generally press hardest on its nearest kindred, and tend to exterminate them.*[17]

The first sentence of this quote explains how more highly evolved individuals within a species will become more plentiful. Consequently, the lesser-evolved members, which could be the parent species or a species that is very closely related, will eventually find complete annihilation, or as Darwin calls it, extinction. In the second sentence, Darwin explains that those most closely related to the more highly evolved species will stand in closest competition.

From a human perspective it means that those who are slightly more evolved will press hardest on their lesser-evolved kin and exterminate them; thus demonstrating the law of natural selection through the inequality of individuals.

Please take into account most of the great wars of mankind. Most have been between closely related races of people, with only slight physical variations between them. Then look at the destruction that each has inflicted upon the other. The Chinese and the Japanese have killed each other in droves throughout history. Though both are strains of the Oriental human—or as Darwin would say, Mongoloids—and they have only slightly different physical adaptations and formations, the competition between them has been fierce.

Look at the two World Wars, when the Germans fought the European nations and the Soviet Union. Again, we see closely allied forms of humans ruthlessly destroying each other. More examples can be found in the constant fighting in the Middle East. The wars between Iraq and Iran have killed thousands of closely related forms.

Some might say that the aggressiveness of humans toward one another is due more to geographical proximity than to physical formation or evolution. In other words, the Japanese and Chinese and the French and Germans fought because they were neighbors. Historically, that is very true. The Germans and Americans, however, were not geographical neighbors, yet they still engaged each other in World War II.

I believe the World Wars signaled a change in the world and technology has been the catalyst for that change. As humans now look to war, and take our instinctive need to press upon and exterminate others, we are no longer bound by geographic location. With technology, miles and oceans no longer limit us. One group of humans can openly war with another group despite being half-a-world apart. Because of that, the lesser-evolved humans are at great risk of being eliminated at a record pace. In the years ahead, we may see the majority of the races/nationalities of mankind eliminated as the human species evolves, and wars break out between the races.

> *We may, I think, assume that the modified descendants of any one species will succeed by so much the better as they become more diversified in structure, and are thus enabled to encroach on places occupied by other beings.*[18]
> — CHARLES DARWIN

39

Therefore, as higher-evolved species move into new territories, there will be an immediate struggle for existence and power. The original inhabitants of the territory, if lesser modified, will eventually reduce in number until an eventual extinction occurs. We observed this as the American Indian receded west upon further colonization by the white man in the early days of the United States. It appears that it is only a matter of time before traces of pure Native Americans are extinguished.

In this modern age, it takes eighteen hours to fly from Atlanta, Georgia, to Johannesburg, South Africa. Just two hundred years ago, it took weeks, at great hardship, to cross the Atlantic via ship; one can now travel from London to New York any day of the week without even taking a nap. This ease of travel has created a modern environment in which all races of people are neighbors! As the superiors multiply and encroach upon the inferiors, there will be friction and eventual loss of life.

> *The enquirer would next come to the important point, whether man tends to increase at so rapid a rate, as to lead to the occasional severe struggle for existence ... Do the races or species of men, whichever term may be applied, encroach on and replace each other, so that some finally become extinct? We shall see that all these questions, as indeed is obvious in respect to most of them, must be answered in the affirmative, in the same manner as with the lower animals.*[19]
>
> — CHARLES DARWIN

Just like the lower animals will act, so, too, shall humans.

This encroachment of humans on other humans is bringing forth an evolutionary fact: when species move into new territories there is a struggle for existence, a struggle for resources, a struggle for power, and a struggle for dominance. Today, there is a massive influx of strains of humans converging upon territories historically occupied by different strains of humans. This encroachment will lead to disharmony. In the end, only the strongest will survive. It is a solid Darwinian conclusion that humans will struggle with their closest kin when territories begin to merge.

> *In the case of varieties of the same species, the struggle will generally be almost equally severe, and we sometimes see the contest soon decided.*[20]
>
> — CHARLES DARWIN

We can see that as one species reproduces successfully, one will fall to eventual extinction. However, please do not feel pity for them or allow religion-based morals to cloud the issue. There is nothing wrong with this outcome. There is nothing wrong with the extinction of the weak. There is nothing wrong with one group of people killing another group of people. It is the way of evolution. Darwin himself explains this situation best:

> *When we reflect on this struggle, we may console ourselves with the full belief, that the war of nature is not incessant, that no fear is felt, that death is generally prompt, and that the vigorous, the healthy, and the happy survive and multiply.*[21]

The war of nature is not perpetuated on the weaker individuals, because for each individual it ends with the certain finality of their own death. As Darwin stated, death is prompt, there is no fear, and in that we should be consoled. As I look into the future and rationalize that one day I might die (barring scientific discoveries), it comforts me to know that I will experience no fear; there is certainly nothing that occurs after death that should be feared.

(I am not sure how Darwin knew there would be no fear upon death because he wrote about it before he died, but I am confident in his genius to rationalize that point through. His ability to make conclusions on things that he never witnessed or experienced was truly remarkable.)

Let's continue.

The law of evolution, which constantly creates more highly evolved and superior beings, will see to it that those who have been modified or improved at a slower rate will naturally suffer most at the hands of the more evolved. This is a constant theme in Darwinian thought:

> *Consequently, each new variety or species, during the progress of its formation, will generally press hardest on its nearest kindred, and tend to exterminate them.*[22]
> — CHARLES DARWIN

Darwin did not claim compassion for the weakest—far from it! He pounded his genius onto paper with the very ink of his pen screaming, *"let the strongest live and the weakest die."*[23]

Any clearer than that and it would be a window. I hope that some of you are starting to see the beauty in a world run by evolution. The stronger should eliminate the weaker for the weaker have no right to life. The human should eliminate the subhuman, and the subhuman should eliminate the nonhuman. This demonstrates the proper relationship of encroachment on a territory and the eventual extermination of the weaker humans. This is sound Darwinian logic.

If it is not clear already, I am on the side of evolution. I am not on the side of conservatism, or liberalism, or toward one philosophy or another. To fully believe in evolution renders all other viewpoints meaningless because the all-encompassing, relentless strength of natural selection allows no other ideology to exist. What I have hoped to show you in the Sub-Law of the Myth of Human Equality is that natural selection, with its unrelenting power to sift populations, leaves no room for any other ideology that creates systems of governance for mankind.

Besides, why should we bother creating other ideologies when we know evolution is law? That has been our problem; the creation of other systems to govern life has caused inefficiencies in the human species. These inefficiencies, which are grossly excessive and present in almost all human races, will be winnowed down by natural selection. That sorting process will come in the form of mass extermination of many humans.

I do have a habit of saving the best for last; this just happens to be my favorite nugget of Darwin dogma. Indeed, we see much of Darwin's personality as he discusses the age-old phenomenon of the relationship between men and women. Before we get started on this topic, please know that I do not dislike women. I like them. In fact, I have one of my own.

So, forget *Men Are From Mars, Women Are From Venus,* here comes Charles Darwin, the love doctor. His views on the relationship between the sexes are dangerously on target. I must confess; there is not much commentary I can make on this subject that would add to the perfectly candid remarks of Darwin, himself. The only question is—and it is one that I have raised before—why is this topic not openly discussed in our society when it is clearly noted in Darwin's scientific writings? Please allow me to share some simple quotes from our founding father:

> *The chief distinction in the intellectual powers of the two sexes is shewn by man attaining to a higher eminence, in whatever he takes up, than woman can attain-whether requiring deep thought, reason, or imagination, or merely the use of the senses and hands. If two lists were made of the most eminent men and women in poetry, painting, sculpture, music, -comprising composition and performance, history, science, and philosophy, with half-a-dozen names under each subject, the two lists would not bear comparison. We may also infer, from the law of the deviation of averages, so well illustrated by Mr. Galton in his work on "Hereditary Genius," that if men are capable of decided eminence over women in many subjects, the average standard of mental power in man must be above that of a woman.*[24]
> — CHARLES DARWIN

Darwin has shown in this first sentence that men can attain a higher position, rank, distinction, or superiority than women can attain in anything they do. This includes tasks that require deep thought, any amount of reason, the vividness of imagination, and anything to do with the use of the senses or hands. Men are not only capable of reaching greater heights with their thoughts; they can accomplish more with their senses as well. According to Darwin, men can see better, hear better, touch truer, and taste more than the simplified version of the human we see in the female—not to mention the fact that men can work better with their hands than can our clumsy feminine counterparts.

Darwin then goes on to say that if a list were created showing the most accomplished men and women covering many disciplines, that the lists would not even bear comparison. Thus, using his own unique criteria, Darwin has proved that women lag far behind men.

He did not stop there. As we conclude the discussion on this quote, Darwin brings it home:

The average standard of mental power in man must be above that of a woman.[25]

— CHARLES DARWIN.

Harvard professor, E.O. Wilson, was right; Darwin is, no doubt, timeless and inspirational. (I wonder why Harvard even bothers admitting women!)

But no, my dear students, this is not the end of his thoughts on the subject. To our great satisfaction, Darwin elaborates further. With the conclusions Darwin just made, how could he not give another rock-solid foundational statement about the intellectual difference between men and woman? I believe Chuck felt the same way (when you know Charles as well as I do, you call him Chuck). To the delight of his staunch followers, he continued to hammer his point home:

Man is more courageous, pugnacious, and energetic than woman, and has more inventive genius.[26]

Is there anything we can conclude from the last statement that differs in any way from what we have already said? I like the notion that men are more courageous, energetic, pugnacious (means "aggressive" for you women out there), and have more inventive genius than do women.

Now, one might think that women have evolved greatly in the last 150 years so that they could have caught up with the men. I am sorry, that is not true. A few generations are not nearly long enough for the entire body of the female population to be lifted to the great heights of the men.

> *In order that woman should reach the same standard as man, she ought, when nearly adult, to be trained to energy and perseverance, and to have her reason and imagination exercised to the highest point; and then she would probably transmit these qualities chiefly to her adult daughters. The whole body of women, however, could not be raised, unless during many generations the women who excelled in the above robust virtues were married, and produced offspring in larger numbers than other women.[27]*
>
> — CHARLES DARWIN

As you can see, it would take wide scale concerted effort on the part of the entire population to train women and raise them in the way of men, and it would take many generations to ingrain those traits throughout the entire female population.

Inquisitive minds out there might just be asking a good question: If women are less evolved are there any similarities between the women of the Caucasian race and men of lower races? Indeed, that would be a good question to ask, for it is reasonable to assume that the women of the white races, who are less-evolved than their male counterparts, might just be on par with the men of the lesser races. Let us turn to Darwin on this question and see if he found any similarities between these seemingly different strains of lesser animals, both in gender and in race:

> *It is generally admitted that with woman the powers of intuition, of rapid perception, and perhaps of imitation, are more strongly marked than in man; but some, at least,*

of these faculties are characteristic of the lower races, and therefore of a past and lower state of civilization.[28]
— CHARLES DARWIN

What shall we conclude then about women? Well, as a whole, the entire body of women is inferior in mental capacity to men. I say this with a humble spirit; I know women do their best, bless their hearts, but the fairer sex, is just that—fair. And since we can't fight science, especially when Darwin is tied to it, women just need take their medicine and take their place.

There is no wonder that Darwin could so easily sum up the value of women when he concluded:

Thus man has ultimately become superior to woman.[29]

My dear disciples, you can do nothing but affirm what Darwin found if you truly understand evolution. The vexing part is why the human race is seemingly frightened to openly talk about this doctrinal truth. This failure, I'm afraid, has led to an overexertion of estrogen-based ideologies and inferiorities in modern society.

Evolution not only works on determining value in race and gender, it works on the sexual appeal of the individual. Now, I promise that we will get into this more in the Sub-Law of Reproduction, but considering that we are discussing human equality, the subject demands we look into the attractiveness of humans here as well. Those of you who are forward thinkers might be able to guess who evolution deems to be the most attractive. If you said people like Charles Darwin, you are correct! Once again, Darwin's theory puts him right at the top of evolutionary status— pure coincidence I am sure.

This will not take long, my eager learners. I assure you that you will enjoy this short, but very informative, lesson on Darwinian evolution and how even our sexual allure is a product of evolution.

First, one must understand a quick fact about Darwin. He was born into an extremely wealthy, elite, and famous family in England. His hefty moneybags swung low, and his silver spoon was stout enough to choke a hippo. On top of that, he was a white male with great evolutionary-enabled intellect. That bit of information will come in handy in understanding this next statement by our founding father and author of our strongest scientific pillars.

> *Many persons are convinced, as it appears to me with justice, that the members of our aristocracy, including under this term all wealthy families in which primogeniture has long prevailed, from having chosen during many generations from all classes the more beautiful women as their wives, have become handsomer, according to the European standard of beauty, than the middle classes; yet the middle classes are placed under equally favourable conditions of life for the perfect development of the body.*[30]
> — CHARLES DARWIN

I must admit, this would certainly have to be one of my favorite Darwin quotes of all time. Charles Darwin stands by and affirms the fact that wealthy families have become physically more desirable than families that hail from lower financial classes. Evolution has created a situation where body types, bone structure, and desirable fleshly shapes are results of one's family having large amounts of money, investments, and other valuable holdings.

Darwin was also quick to point out that the conditions of life are equally good for the middle class as they are for the wealthy. This confirms that only the financial rank is responsible for the perfectly developed body. Otherwise, one might assume that financial prosperity would give such comforts and luxury that the mere lifestyle difference between the rich and the poor would result in a better-looking individual.

For an example, one might argue that a rich person would have a classier, more attractive presentation of their body because of their clothing, make-up, and the adornment of fine jewelry; whereas a peasant's appearance may be tattered and scarred by harsher labor environments and be covered in torn and odorous clothing. The poor would also embody the empty stares that vagrants usually have from living in huddled and diseased communities and such would mar an otherwise decent looking fellow. Therefore, the factors that impact physical appearance would be neither genetic, nor the results of evolution, but rather the conditions of life.

But according to Darwin, these factors have not led to the wide disparity of physical beauty between the classes. Darwin argued that the living conditions are equally good for both classes, so the genetic pressures of evolution, driven by unique choices afforded to the upwardly affluent, have had the impact needed to create more attractive human bodies for the wealthy. Evidently, evolution likes a good asset.

If we look closely at Darwin at this point, we find an intriguing observation, and the coincidence of that observation is most amusing. The theory he penned not only described him as a member of the highest, most "favored" race and of the highest gender, but also argued he was among those with the highest sexual appeal because he had been born into a wealthy family. How nice for him.

The evidence continues to mount that Charles Darwin had rock-solid sensibility, complete soundness of mind, and a healthy self-image—all attributes that led to his discoveries that men came from apes, that people of darker skin tones are subhuman (or not human at all), that woman are stupid, and that he was a member of the best looking population, of the most evolved race, of the most evolved gender.

It was this man's complete sanity of mind that shaped the modern scientific movement known as evolution.

Many of you with morally throttled minds have no doubt encountered "moral hurdles" with the Sub-Law of the Myth of Human Equality. The indoctrination of religion has caused many to fall away from the instincts that have made our species so strong. So without further ado, it is time, once and for all, to end the religious debacle by presenting the Sub-Law of the Doctrine of Sin.

ENDNOTES

1   http://www.reefed.edu.au/home/glossary.

THE SUB-LAW OF THE MYTH OF HUMAN EQUALITY

2   Darwin, Charles, *From So Simple a Beginning*, New York: W.W. Norton & Company, Inc., 2006, pg. 490.

3   Ibid., pg. 209.

4   Ibid., pg. 209.

5   Ibid., pg. 1419.

6   Ibid., pg. 209.

7   Ibid., pg. 209.

8   Ibid., pg. 196.

9   Ibid., pg. 196.

10 Ibid., pg. 199.

11 Ibid., pg. 193.

12 Ibid., pg. 783.

13 Ibid., pg. 11.

14 Ibid., pg. 798

15 Ibid., pg. 489.

16 Ibid., pg. 489.

17 Ibid., pg. 520.

18 Ibid., pg. 524.

19 Ibid., pg. 784.

20 Ibid., pg. 498.

21 Ibid., pg. 500.

22 Ibid., pg. 520.

23 Ibid., pg. 605.

24 Ibid., pg. 1204.

25 Ibid., pg. 1204.

26 Ibid., pg. 1198.

27 Ibid., pg. 1205.

28 Ibid., pg. 1203.

29 Ibid., pg. 1205.

30 Ibid., pg. 1220.

CHAPTER SIX

THE SUB LAW OF THE DOCTRINE OF SIN

From the Cellar of Dr. Iman Oxidant

MY FAITHLESS multitudes:

Humanity is plagued by a delusion that is unique to our species. No other animal on the planet has created a doctrine of foolishness like that which mankind has produced.

I suppose any species that reaches the intellectual heights modern man has reached might start to wonder about its origins. It may seem natural to search for some other meaning beyond the physical world; one that cannot be seen, tasted, or felt.

> *Nor is it difficult to comprehend how* [religion] *arose. As soon as the important faculties of the imagination, wonder, and curiosity, together with some power of reasoning, had become partially developed, man would naturally have craved to understand what was passing around him, and have vaguely speculated on his own existence.* [1]
> — CHARLES DARWIN

I can understand how early man became curious about their origins; they did not have the intellectual storehouses of knowledge we have today. Their adoption and development of make-believe creation stories is a cute testimony to the simplistic mind our species once had. It is because of that simplicity, however, that we now need to change diapers, and discuss the invention of religion and the subsequent slavery to sin.

My faithful reader, our biggest battle in the war over the minds of man is in the very trenches of religion. We now stand at the impasse of religion vs. evolution. The two cannot coexist; eventually one will be thoroughly defeated at the hands of the other. Today, we are on the front lines of that conflict, and Darwin is our general.

In this war, we know religion asserts that a god is the creator and designer of this complex life. But even more shocking is the Christian claim that humans are naturally flawed and need the approval and guidance of an authoritative god. This belief is embodied in the Christian view of sin.

In contrast, evolution is the study of life in which we admit we see no evidence of god as a creator or designer; "Wheels have designers, but evolution does not."

This is a critical point of differentiation, because if the human species had a designer, that would denote thought, purpose, and planning in our existence, which is not the case. The human species has no designer's purpose, we are not a product of careful thought, and we are not a planned existence. We exist by sheer chance.

> *Humanity was thus born of Earth. However elevated in power over the rest of life, however exalted in self-image, we were descended from animals by the same blind force that created those animals, and we remain a member of species of this planet's biosphere.*[2]
>
> — E.O. WILSON

Religion adheres to an assertion that is the polar opposite of evolution! Christianity, in particular, states that humanity was born in the image of the same god who breathed life into Adam, who was allegedly the original human.

The only way evolution and religion can co-exist is if one asserts that religion is merely part of our history, and it is through that history of religion that we can learn about the growth and evolution of man.

> *Creation myths were in a sense the beginning of science itself. Fabricating them was the best the early scribes could do to explain the universe and human existence. Yet the high risk is the ease with which alliances between religions and tribalism are made. Then comes bigotry and the dehumanization of infidels.*[3]
>
> — E.O. WILSON

I understand what well-intentioned E.O. Wilson was trying to accomplish with his statement; he was attempting to sway the masses toward evolution by using emotional triggers such as bigotry and then tie those culturally frowned-upon sentiments to religion.

But that is not being intellectually honest, and the pursuit upon which we have embarked is all about evolutionary honesty! I promise you; if men of reason swim in those pools of ethics and morality, we will be swimming with *Jaws*.

Evolutionary scientists and intellectuals believe in natural selection, survival of the fittest, and the iron fist of evolution. We don't adhere to a code of moral conduct that was derived from mythology. The idea of bigotry is not even on our mainframe; we deal only with the factual implications of higher races versus lower races, and we do not filter those scientific facts through any manufactured moral code.

Much like feeling the emotions of a poet through his writings or the music of a musician as he plays, the historical use of religion in understanding the moods of past civilizations can be useful. But far be it from me to claim that the role of religion is any more important than just a few colorful pages of human history. The truth is that we are products of evolution, not the creation of a potter, molder, designer—or a god.

Therefore, evolution and modern religion are two separate paradigms that attempt to explain our origins, and as you will see, they cannot, and do not coexist. This is an important Sub-Law, mind you, because too often I have heard of people who believe in the science of evolution (rightfully so), yet still misguidedly subscribe to Christianity or some other religion that aspires to reach god.

For me, nothing is more frustrating. Evolutionists cannot believe in god, especially in the god of Christianity. Having ill-educated evolutionists still hanging on to the primitive notions of religion frustrates our whole movement and impedes our progress. As we unravel the secrets of Darwin, I will show the entire world that the beliefs that accompany religion are the biggest threats to evolution, and therefore, the biggest threats to our civilization.

My attentive, purposeless pupils, I introduce you to enlightenment, to freedom, to pure Darwinian instinct. Welcome to the Sub-Law of the Doctrine of Sin.

We are going to eliminate religious dogma from our thinking and lifestyles. Freeing ourselves from religion will open us up to a new form of dominance on this planet, and consequently, prevent our extinction by eliminating inefficiencies that stem from this ideological belief system.

When the plank of Christianity is burned, all other religions will go up in flames as well. There is no doubt Darwin saw Christianity as the 800-pound gorilla. When he named his diagram, the "Tree of Life," which describes the evolution of organisms, he deliberately fired a warning shot across the bow of the Christian ship, as there is also a "Tree of Life" in the Book of Genesis's Garden of Eden. In his superior intellect, Darwin left a trail of breadcrumbs so we could find the true enemy of evolution.

Darwin does not leave any cloudy issue with regard to where he stands:

> *I can indeed hardly see how anyone ought to wish Christianity to be true; for if so the plain language of the text seems to show that the men who do not believe, and this would include my father, brother and almost all my best friends, will be everlastingly punished. And that is a damnable doctrine.*[4]

In mounting our case against Christianity and sin, authenticity is paramount. As we hold true to Darwin, we need to hold true to the Bible in the same respect. This authenticity will please you, for I am sure this Sub-Law will write the final chapter on Christianity. So, persevere with me through our dissection of Christianity, for the end will certainly justify the means.

The Christian religion is an outdated belief system created by men who have now been out-evolved over the course of many generations. The Bible has caused too many problems for the world already. For this reason, we are about to cut a wide swath through the heart of Christianity. Be prepared to bow before Darwin's evolution all you religious zealots!

I know you have waited patiently for this moment. You see, my good listeners, with my highly evolved intelligence, I have an ability to reason that very few people on the planet can ever acquire. I have simplified the argument for you; watering it down for those who are intellectually dry.

At its core, Christianity maintains the belief that something called sin causes failure in the human race. The *International Standard Bible Encyclopedia* explains it like this:

> *A fairly exact definition of sin based on biblical data would be that sin is the transgression of the law of God. (See 1 John 3:4.) Ordinarily, sin is defined simply as "the transgression of the law," but the idea of God is so completely the essential conception of the entire biblical revelation that we can best define sin as disobedience to the law of God.*[5]

It is this sin that creates a chasm between man and god. One could say that the problem of sin is the overwhelming theme of the Bible and the solution to that problem is Jesus—the believers' "Savior" who saves them from their sins. To put it into layman terms, the Bible claims that if sin is the sickness, then Jesus is the cure.

The predominant Christian assumption is that sin leads to eternal damnation in hell. And because of this potentially frightening end, there is much anxiety about the eternal destination of the human soul (there is no scientific evidence for a soul, by the way).

It is because of this fear that the Bible has been so successful in moving people into action. It creates and defines a problem for the human species and calls it sin. This sin has severe consequences—damnation in a lake of fire for eternity, complete with plenty of weeping and gnashing of teeth (a little dramatic don't you think?).

Christianity baits the individual through the terrifying concept of damnation and then offers a surefire way out called salvation. It is a classic fear tactic; create a problem and then solve it. This strategy has been quite successful in replicating itself over and over again throughout many generations and many cultures. Fortunately for the world, evolution is here to end it.

When people succumb to the concept that sin exists and that it has consequences, they will naturally wonder if they are sinners. The Bible is quick to point out that yes, they are! The Christian viewpoint is that sin entraps *every* human being in the world—even your very own Dr. Iman Oxidant!

For all have sinned and fall short of the glory of God.
— ROMANS 3:23

Therefore, by default, everyone is deemed guilty of sin, which puts us all on a one-way highway to hell.

In my opinion, that is where the Bible gets greedy. Trying to include everyone into its belief system is a lesson in futility. Having the arrogance to claim everyone is a sinner is atrocious and having the gall to claim that all must abide by this belief is absurd. Then to add that forgiveness of these sins can only come through one person, let alone a carpenter's son, Jesus of Nazareth, is not even humanly rational.

Whoever believes in him is not condemned, but whoever does not believe is condemned already, because he has not believed in the name of the only Son of God.
— JOHN 3:18

It angers me when Christians judge and condemn me and try to throw their belief system onto me! Who are *they* to define what is right? I am above sin, I am not bound by its rules and guidelines—and I have evolution to prove it! Christians, and their fundamentalist ways, are becoming very dangerous. Because of that I am more eager than ever to disprove them.

Due to the Bible's assertion that sin is something that plagues the entire world, Jesus Christ is revered by His followers as the Savior of the world because he can forgive their sin.

> *For God so loved the world, that he gave his only Son, that whoever believes in him should not perish but have eternal life. For God did not send his Son into the world to condemn the world, but in order that the world might be saved through him.*
>
> — John 3:16-17

This is where most debates break down between Christians and evolutionists; the argument over the divinity of Jesus Christ. I am not your average evolutionist, however. I think deeper, my eyes are clearer, and my wit is sharper. I will not stop where others have faded. With my far-reaching abilities, I understand that sometimes a simple question can cause a deeper quake than a lengthy thesis.

So without further ado, under the direction of Darwin, let us shatter the foundations of these under-evolved humans, and reduce the Bible to what it should be, a book of stories with less meaning than Greek mythology. Through this Sub-Law, I will reinforce every textbook that teaches mankind's ancestor was an ape-like creature who once sniffed the ground with as much sense as a dog returning to its vomit.

With that glorious history—from which I am proud to descend—I ask you: *What if there is no sin?*

What if it could be proven that sin does not exist? If there is no sin there would be no need for redemption from that sin. In other words, if there is no sickness there is no need for a cure. What an intoxicating thought, no?

According to the Bible, the presence of sin created the whole need for Jesus. Keen deductive reasoning shows us that if there were no sin in the world, there would be no reason for forgiveness from that sin. If there is no sin, then there is no fear of punishment as a result of sin. And if there is no punishment for sin, then there could be no hell.

Likewise, there can be no reward for receiving forgiveness of those sins. This "promised reward," a life filled with peace, joy, and happiness for all eternity in the Kingdom of Heaven, is moot, because one cannot accept a reward for something that does not exist. Therefore, Jesus, whose purpose was to conquer sin, finds himself purposeless. There is no hope of eternal life for mankind and Jesus cannot grant a gateway to a false heaven through forgiving imaginary sins.

Jesus promised a blissful eternal existence full of joy without fear, life without death, gain where there is no loss, and health without a trace of sickness. This is a false promise, and knowing it is false is of high value. Why entertain hope in an empty promise, no matter how incredible it seems to be? It just creates inefficiencies in our actions. To believe there is life after death allows us to live in a world of lies; lies that cause us to be compassionate toward some humans, when in fact, we should exterminate them.

In reality, it is best for us to come to terms with what we are—an accident of immense proportions. We must fight, scratch, scramble, and spit through the short time we have here on Earth using survival of the fittest tactics to navigate our way. We are born of earth, not of god. This earth, this life, the blood that runs in our veins and the basic urges of our instinct are all we have!

On that note, I reiterate our foundational truth:

Let the strongest live and weakest die.[6]
— CHARLES DARWIN

This is in sharp contrast to the Biblical view:

You shall love your neighbor as yourself.
— MARK 12:31

The two ideologies compete at every philosophical front. The manners in which Darwin and Jesus viewed the world and its inhabitants could not be any further apart. Anyone who believes both of these doctrines at the same time has a mind divided against itself. Christianity and evolution, at their core foundations, not only disagree, but actually fight to destroy one another.

Whereas Jesus came to heal and save the weak and weary, Darwin convinced us that the strong should press upon and kill the weak; that is the only way to make a species stronger and enable the process of evolution to work.

In light of this, we will remove Jesus from our lives. Evolutionary thought creates a world free from guilt, free from fault, free from responsibility, free from authority and free from an outdated set of morals and ethics.

On with the sin crusade!

The definition of sin we used earlier was quite revealing. It taught us that sin is not breaking the laws of man; it is breaking the laws of God. The laws of God are outlined in the Ten Commandments, which came about through an alleged divine encounter between Moses and God on Mount Sinai.

> *And he gave to Moses, when he had finished speaking with him on Mount Sinai, the two tablets of the testimony, tablets of stone, written with the finger of God.*
> — EXODUS 31:18

The definition of sin did not end there, however. Sin eventually became identified as anything that is contrary to the teachings of the Bible—homosexuality, greed, and pride for example. Sin is, therefore, closely linked to supernatural intervention and interpretation because it shows no proof of origination or purpose outside of the realm of God. If God did not intervene and give the Ten Commandments to Moses, we could not even define sin, because no transgression of the law of God could be identified.

Another question of interest regards where sin came from. This is not quite the same thing as how it is identified. We already see that sin is identified through the transgression of God's law, but identifying something and creating it are two different things.

Christian thought teaches us that sin came into the world through Adam, the infamous first human, who was formed from the dust of the ground. God himself blew life into this being and called him man—or so the story goes.

> *Then the Lord God formed the man of dust from the ground and breathed into his nostrils the breath of life, and the man became a living creature. And the Lord God planted a garden in Eden, in the east, and there he put the man whom he had formed.*
> — GENESIS 2:7-8

Understanding how the Bible views the creation of man is critical. We see through this creation story the tale of how sin came into the world:

> *Students of all schools are agreed that we have in the Old Testament story of the fall of Adam an eternally true account of the way sin comes into the world.*[7]
> —INTERNATIONAL STANDARD BIBLE ENCYCLOPEDIA

It has just been asserted that the entire Christian school of thought is founded on the belief that sin came into the world as a result of the disobedience of Adam.

> *And to Adam [God] said, "Because you have listened to the voice of your wife and have eaten of the tree of which I commanded you, 'You shall not eat of it,' cursed is the ground because of you; in pain you shall eat of it all the days of your life."*
> — GENESIS 3:17

To understand sin and what it means in Christianity, we need to understand another platform of thought that is inseparable from Christianity. When sin entered the world due to Adam's disobedience to God, it created a separation between God and mankind. No longer could man please God, because he was now under the "curse" of Adam, a curse called sin. The result of that sin is catastrophic; the Bible clearly states in Romans 6:23 that "the wages of sin is death."

The term "wages" was not used lightly here. Wages, by their nature, are something that must be paid. Wages are also paid after the fact; a person would typically work for a day's wages and at the end of the day he would be paid for his work.

The Bible explains that sin results in death and that sin has already been committed by the human race as a whole. This blanketing style of inclusion falls in line with the thought that "all have sinned and fall short of the glory of God" doctrine, of which we spoke earlier.

In the account of Genesis, we see the creation of sin at the hands of Adam. If mankind is not to live in eternal damnation, according to the Bible, there needs to be salvation from the sin that Adam introduced to the world. It is at this point, that the Bible creates the need for a savior. This need eventually was manifested with the rise of Jesus.

> The next day he saw Jesus coming toward him, and said, "Behold, the Lamb of God, who takes away the sin of the world!"
>
> — JOHN 1:29

Through this verse we quickly see that Jesus is positioned in the New Testament to be the one who can release the world from the sin that was introduced to the human race through Adam. Jesus is the cure for the sin; Jesus fixes the problem Adam created.

> For as in Adam all die, so also in Christ shall all be made alive.
>
> — 1 CORINTHIANS 15:22

Now that we understand the relationship between Adam and Jesus, I can make good on my promise to remove sin from the world and finally rid the human race of the need for Jesus, salvation, heaven, and certainly hell.

To dispel sin, we need only dispel the authenticity of the Genesis creation story. Of course, that is something easily done; that is one area our scientists have shown their quality. However, for the novices under my tutelage, let me briefly sum up the argument for you. We are going to put biblical and Darwinian thought side-by-side to illustrate to the ignorant why the Genesis creation story cannot be taken seriously.

> *And God said, "Let the earth bring forth living creatures according to their kinds—livestock and creeping things and beasts of the earth according to their kinds." And it was so. And God made the beasts of the earth according to their kinds and the livestock according to their kinds, and everything that creeps on the ground according to its kind. And God saw that it was good.*
>
> — GENESIS 1:24-25

In these verses we see a sharp contrast between biblical and evolutionary thought. The most interesting aspect of the creation account is that it insinuates that God created all the land creatures "according to their kinds," clearly separating the origins of individual groups of organisms. This is in sharp violation of our law of evolution! Darwin is clear when he states:

> *I can entertain no doubt, after the most deliberate study and dispassionate judgment of which I am capable, that the view which most naturalists entertain, and which I formerly entertained—namely, that each species has been independently created—is erroneous.*[8]
>
> — CHARLES DARWIN

Darwin states that the idea—which is propagated by the Bible—that each species is created independently, or by their kind, is erroneous. Evolution teaches that given enough time, animals diverge in character and new species are formed.

> *Then God said, "Let us make man in our image, after*
> *our likeness. And let them have dominion over the fish*
> *of the sea and over the birds of the heavens and over the*
> *livestock and over all the earth and over every creeping*
> *thing that creeps on the earth."*
>
> *So God created man in his own image, in the image of*
> *God he created him; male and female he created them.*
> — GENESIS 1:26-27

Here we see the creation of the first man, Adam (again within a day, so the story goes). Man is made in the image of God. The Bible is absolutely clear that man is a separate creation from the rest of the organisms on the planet. Man was somehow set aside and given authority, or dominion, over the Earth because of some appointed order set in motion by God.

In the next chapter of Genesis, the Bible goes into greater detail about how god created man:

> *Then the Lord God formed the man of dust from the*
> *ground and breathed into his nostrils the breath of life, and*
> *the man became a living creature.*
> — GENESIS 2:7

This verse puts to rest any notion that Adam was a product of ancestry. Genesis says the first man was formed out of dust, not a product of billions of years of evolution. Darwin sharply disputes this Genesis story:

> *On the view that each species has been independently*
> *created, with all its parts as we now see them, I can see no*
> *explanation.*[9]

Furthermore, Darwin states that mankind was not created as a stand-alone creation in the image of god. Instead, mankind is related, through common ancestry, to every other organism on the planet:

A corollary of the highest importance may be deduced from the foregoing remarks, namely, that the structure of every organic being is related.[10]

And finally, for those who need to have the dead horse beaten, Darwin explicitly contradicted the Bible on the specific origin of man:

Man, as I have attempted to shew, is certainly descended from some ape-like creature.[11]

There can be no dispute of these Darwinian thoughts; these foundational statements are the facts of evolution to which the entire educated world now adheres. I have shown that evolutionary theory contradicts the story of creation found in Genesis.

Humans are descended from an ape-like creature, which is certainly not a god. And therefore, there cannot be a real historical figure of Adam who was formed out of the dust in the image of God. To stay on point, it almost goes without saying that Eve could not have borrowed a rib from a non-existent Adam to be formed. And that, I am afraid, makes the poor snake charmer's existence completely impossible.

It is time for the climax. This is where we tie all the loose ends together and the shards of glass are assembled to create the mosaic masterpiece Darwin has created.

As we learned earlier, throughout the very verses of the Bible, Adam and Jesus Christ are completely inseparable; Jesus' teachings were hinged on the biblical thought that Adam was created as the first man and brought sin into the world, and if that sin was not defeated, the entire human race would be eternally guilty and be sentenced to eternal damnation in hell.

According to evolutionary thought, however, we have learned that the Genesis creation story, including Adam as the first man, is nothing more than a myth. That evolutionary conclusion leads to a string of failures for the Bible.

Consequently, there could not have been a transgression at the hands of Adam that would have introduced sin into the world. If there is no sin in the world, then there is no judgment for sin, no penalty for sin, and no reward for being forgiven of that sin.

With this evident absence of sin, there is no need for salvation. Therefore, there is no need for a savior and ultimately no practical need for Jesus Christ. The message, purpose, and life of Jesus of Nazareth is completely extinguished and rendered useless because of the core doctrines of evolution.

The Sub-Law of the Doctrine of Sin has allowed us to be freed from sin and know that our freedom comes from the strength of human reason! Our reason has no error in it; for we have based it on Darwin and his evolutionary thought.

My fellow humans, in the absence of Christianity and sin, we are freed from the fundamentalist's moral high ground. Indeed, the concepts of monogamy, marriage, love your neighbor as yourself, and turn the other cheek are erased as valid lifestyle options.

If there is no sin, then there is no ultimate authority that determines right and wrong. Our parameters of right and wrong are now defined by what benefits us in evolutionary terms as individuals and as a species. Our parameters of action, based on the Doctrine of Sin, state that moral authority does not bind us. Instead, we are granted a natural authority, which gives us the right to perpetuate our species in any way we are strong enough to see fit.

Welcome to the world without sin courtesy of Charles Darwin's evolution!

ENDNOTES

1   Darwin, Charles, *From So Simple a Beginning*, New York: W.W. Norton & Company, Inc., 2006, pg. 815.

2   Ibid., pg. 12.

3   Ibid., pg. 1483.

4   Ibid., pg. 1482.

5 *International Standard Bible Encyclopedia*, http://www.studylight. org/enc/isb/view.cgi?number=T8184.

6 Darwin, Charles, *From So Simple a Beginning,* New York: W.W. Norton & Company, Inc., 2006, pg. 605.

7 *International Standard Bible Encyclopedia*, http://www.studylight. org/enc/isb/view.cgi?number=T8184.

8 Darwin, Charles, *From So Simple a Beginning,* New York: W.W. Norton & Company, Inc., 2006, pg. 452.

9 Ibid., pg. 547.

10 Ibid., pg. 498.

11 Ibid., pg. 1223.

CHAPTER SEVEN

THE SUB-LAW OF REPRODUCTION

From the cellar of Dr. Iman Oxidant

MY EAGER breeders:

The gloves are coming off. We have reached the point where we have finally stripped ourselves of Christianity, sin, and morality. We no longer have interference with our human reason and have nothing to be ashamed of when presenting pure evolutionary logic. In fact, we should have unceasing candor, honesty, and accuracy.

So, my accidentally and chaotically-formed humans, let us dive in and understand that this next Sub-Law promises to be just as revolutionary as the first two.

We will begin this discussion with a simple refresher on natural selection—individuals who are more suited to their environment will tend to survive and produce more offspring.

For the purposes of this Sub-Law, we will discuss the second part of this statement: *produce more offspring.* The Sub-Law of Reproduction is concerned with the reproduction of the human species—or the procreation of offspring.

Some might say this is the most important Sub-Law—a claim that is hard to argue against. Reproduction is the single most important purpose of a species. Without successful breeding, a species will become extinct.

It is at this juncture we see a dilemma with humanity. Humans are at odds with the principles of evolution because of the sexual habits we maintain.

What has endangered our reproductive habits in recent times are the overriding Christian and religious views on sex, marriage, and adultery. Since we now understand that Christianity and sin are dead, the Commandments "You shall not commit adultery" and "You shall not covet your neighbor's wife" no longer apply. Therefore, to allow Christianity to introduce monogamy and marriage as acceptable lifestyles within the human species was a horrible mistake; a mistake with which we no longer have to live.

Marriage and monogamous sexual relationships are not ideologies that are associated with nature or the ways of evolution. Consequently, they have no place in our modern society. In fact, sexuality between male and female humans is completely different when you remove Christianity and sin from the equation and focus instead on the ways of nature and evolution.

Today, we will discuss proper sexual activity. Evolution has granted us the right to redefine our sex lives. Darwin taught us about evolutionary reproduction through his explanation of "Sexual Selection."

> And this leads me to say a few words on what I call Sexual Selection. This depends, not on a struggle for existence, but on a struggle between the males for possession of the females; the result is not death to the unsuccessful competitor, but few or no offspring.[1]

We see that sexual activity is not a sequence of events during which the man and the woman find each other attractive and reproduction occurs within a mutually agreed upon framework.

In Darwin's words, there is nothing "mutual" about reproduction; it is about strength, right, and might. Darwin states that women, who are desired by men, will be bred, whether or not the woman feels inclined to be bred by that particular man.

We learned in the Sub-Law of the Myth of Human Equality that women are inferior to men in every manner and discipline. So under what basis could a woman, with a more feeble mind than her potential male breeder, rationalize if she should be bred or not? She can't. It is the male's decision. Evolution relies on the superior reasoning of the male to make that judgment and then act upon it.

That being said, there are timeframes within the feminine cycle where a woman is more inclined to allow herself to be bred by a man. Nature has inflamed her lustful desires during this time because there are specific points in the woman's cycle where it is most optimal for her to be impregnated and it is imperative for the sake of the species, that she is sexually engaged during that short window of time. Just like her relative, the cow.

When a cow is in heat, it will easily allow the bull to couple with it; if she is not in heat, she will try to prevent the bull from mounting her. But not just any bull can mount the cow. The bull that defeats other competing bulls has the right to pass on its genetics and breed the cow. This creates an efficient reproductive environment—one that humans must emulate.

A dominant species, like humans, must have become dominant because the species carried out the doctrine of "Survival of the Fittest" more efficiently than all the other life forms that could have occupied the same niche. In other words, the human had stronger evolutionary development. The hand of natural selection is always ready and willing to accept and reject individuals based upon their strengths and weaknesses.

As Darwin said:

All organic beings are exposed to severe competition.[2]

This struggle for life allows the strongest to continue to procreate and the weakest to die with few or no offspring. This leads to the continuation of some species and the extinction of others. Up until this point, humans have been the victor under this set of immovable circumstances. Therefore, modern humans are descendents of the prehuman species that won this evolutionary battle.

But here is our dilemma and the trigger that sprung these Sub-Laws into action: humans have turned away from the very habits that made us strong, and this epidemic is species-wide! We have stopped engaging in the natural power struggle because of morality and the softness of Christianity. This makes us weak by evolutionary standards and will eventually lead to our extinction; we have buried our instincts in piles of morality and political correctness.

Darwin teaches:

> [V]*arieties, in order to become in any degree permanent, necessarily have to struggle with the other inhabitants of the country, the species which are already dominant will be the most likely to yield offspring which, though in some slight degree modified, will still inherit those advantages that enabled their parents to become dominant over their compatriots.*[3]

With this knowledge, we can easily rationalize that if a species knowingly rejects the laws of nature—like survival of the fittest, natural selection, and sexual selection—that very species, which was initially strong, will become weak. Weakness leads to elimination.

In contrast, if no repercussions resulted from rejecting the ways of evolution, that would insinuate that it was not the law of the land because it was not strong enough to render judgment. But evolution is real; every scientist, teacher, and person of reason accepts that it is real. Evolution is the very fiber of our lives and the science that explains our very existence.

> *The concept of evolution is inextricable from the language of all life sciences.*[4]
> — THE AMERICAN SOCIETY FOR CELL BIOLOGY

Here is the danger: somewhere in the last few thousand years we have stopped living according to the design of evolution and stopped breeding with strength in mind. There is no doubt that this change can be attributed to the fallacy of Christianity, which fosters the concepts of compassion, morals, love, and an eternal soul—all of which are absent in the animal kingdom. There is no soul seen in the makeup of DNA; lions do not consider right and wrong, and rhinos do not fall in love. Evolution proves that we are neither bound by morals, nor given false hope with the notion of an eternal soul. Humans are the only animals that have ascribed to such ideologies.

Continued belief in such myths will certainly bring destruction. Quite frankly, I am surprised we have existed this long when we have so arrogantly disobeyed the ways of nature.

But there is a bigger problem. If the human species shows a continued adherence to these Christian notions, that would suggest a design and purpose bestowed upon us. Such a design would separate us from the rest of the animal kingdom.

Let me be painfully clear. Any attempt to present humans as anything more than simply the dominant species on this planet will be catastrophic. Divine design also brings with it the illusion of morals, sacrifice, compassion, and love, none of which can be justified through scientific research. There is no "morality chromosome" or "Good Samaritan" instinct in any animal in the wild. As we all know, an injured animal is destroyed in the wild, not helped. Therefore, since we are also inhabitants of this wild planet, we must act like it—or face questions we don't want to answer.

Morality is simply intelligence run awry. In the future, as our intelligence gets over the hump of religion, we will cease to feel the need for morals and ethics. Only argument, science, and displays of strength will determine who controls resources and breeding rights, and who has a right to live or will be ushered unto death. As the lion hunts and kills, so do the shark and the snake, and so shall Man.

Using the Darwinian principle that the human is just another species within the animal kingdom, our sexual actions should more closely resemble those of other animals. In fact, because we have the ability to learn, we should study the breeding habits of all the species of the world and determine which ones make the most sense for us to adopt for ourselves. Their instincts have not deteriorated like those of humans and they have not been stained by religion. Animals are more correct in an evolutionary sense than are we.

It is clear we have been stupefied by our own imaginations. Modern society has been caught in a web that needs to be unwound. It is time for evolutionary logic to be implemented in our lives; it has the power to unravel the very fabric of our morality-based society.

Human breeding should shift toward the Darwinian practices regarding the basic struggle for life, and steer clear of cultural restraints like abstinence, which is the suffocation of our instinct, and marriage, which is the caging of our possibilities. Women should have no role in this struggle. Breeding is not a matter of a man wooing a woman and winning her heart; indeed, it is easier for a man if the woman tosses the ballot his way, but that does not matter to evolution. It is between men only and is decided when one man is victorious over another. The victor reaps the spoils, in this case the woman.

This model is available for us to view in nature. When a male peacock flaunts himself to attract a female, mating is still subject to another male challenging him. The female peacock may come looking to mate, as she will make herself available to be bred because of her instinct, but the female waits until there is a victor in the battle. It is not her decision; she would not deny the victorious male whoever that may be. In the end, this competition leads to stronger offspring:

> Generally, the most vigorous males, those which are best fitted for their places in nature, will leave most progeny. But in many cases, victory will depend not on general vigour, but on having special weapons, confined to the male sex.[5]
>
> — CHARLES DARWIN

Here we see the inevitable and obviously aggressive side of nature. Darwin explains a seemingly physical confrontation between males; either the physically strongest male or the one with the best weaponry shall decide who leaves the most progeny. The male who is most powerful will be victorious. This is just further evidence that Christianity has ruined our mating practices by introducing the notion that there is equality between men and women:

> *For the wife does not have authority over her own body, but the husband does. Likewise the husband does not have authority over his own body, but the wife does.*
> — I CORINTHIANS 7:4

Here we see another ludicrous position of Christianity in which the man and woman, within a marriage relationship, have authority over each other's bodies—thereby establishing a sense of equality. Obviously, in evolutionary thought, the man has authority over the body of the woman, but the woman certainly does not have authority over the man.

Furthermore, this verse is incomplete without the addition of this other peculiar Bible verse:

> *Therefore a man shall leave his father and mother and hold fast to his wife, and the two shall become one flesh.*
> — EPHESIANS 5:31

The Christian notion of love between one man and one woman becomes completely bizarre when reading this verse. The Bible asserts that through marriage, a man and woman actually become one flesh. Honestly, I don't even know what that means! Evidently, it is suggesting a bond that is unbreakable. According to our evolutionary understanding, this is one of the most ridiculous philosophies that has ever crossed my desk.

Under biblical constraints, one could easily see how a husband would sacrifice his very life for his wife. It goes without saying, that such self-sacrifice is neither expected nor endorsed in evolutionary thought—indeed, it is a sign of weakness.

Evolution is a completely selfish, self-centered, and self-perpetuating science. It *must* be for the individual to leave strong offspring and for the subsequent strengthening of the species. Evolution is a science of the individual and how the individual must fight to stay alive. It is about a man effectively breeding the women of his choice so he can have offspring and thereby secure his line of DNA.

Conversely, Christian marriage is almost self-deprecating. The Bible is so clear in this self-sacrificial philosophy, in fact, that the New Testament shows that Jesus sacrificed his life for all mankind because there is no greater display of love than to give your life for another.

This attitude of sacrifice is further illustrated in the lives—and deaths—of several other main characters in the New Testament. Nearly all of the disciples were imprisoned, killed, or both for the "sake of the gospel." That is sheer lunacy; no earthly gain can be seen from these men allowing their short, completely accidental lives to be destroyed for the cause of Christianity. This willful destruction of one's own body has no business in nature. What could have possibly been their motive for doing such a thing? Another mystery of the Bible, I suppose.

Simply put, if you sacrifice your own life to save another you are frail. The apostles were weak and Jesus was unsuitable for survival as well because he died at the young age of thirty-three without leaving any offspring. Darwin taught that the strong shall live and the weak should die. If you are the one sacrificing yourself to save others, then you die and are therefore counted as feeble.

I hope you see how far we have fallen as a species by allowing Christianity to influence our social parameters—we have been taught to suppress the aggressive sexual instincts of the animal in all of us.

Our strength is found in science and the law of evolution. Once we allow thoughts of morality to stand on equal footing with science, we are weakened. There is no safe escape from the Pandora's box that has been opened. The continued existence of morality is about as likely as a male lion willingly sharing his prey with a starving hyena.

Remember, humans are not special and are not designed for some lofty purpose. Therefore, we cannot expect to be treated differently by nature than any other animal. Let the strongest live and the weakest die!

Furthermore, when the law of evolution is actually followed in human mating practices, the legal courts of many nations convict and incarcerate the male who took his rightful place in nature and followed the Sexual Selection technique promoted by Darwin.

If a man is strong enough to possess a woman whom he finds desirable and attractive, he should copulate with her whether she likes it or not and local bodies of government should have no right to interfere with nature. How can we teach the principles of evolution to children in our publicly funded schools, and then criminalize them when they apply the truth of that science to their lives and act upon those principles? It is utterly asinine.

It is also important to interject that strength is determined by the demands of the environment. A man who can bench press 350 pounds will probably not out-duel a frail man who is wielding a fire arm. So strength can be defined in many ways, but it is most importantly defined as he who possesses what is necessary to win. So whether one is strong mentally, or one is strong physically, each unique environment and competition will decide who is strong, what is strong, who is weak, and what is weak.

If we are successful in our teaching, we should see evolutionary lifestyles manifest themselves in humans as they grow older. These lifestyles should be encouraged and not punished. Males who take what they want are only acting within the ways of evolution; there can be no crime in that! If evolution is completely true, as we have concluded, then the actions and behavior of those following its principles are only natural and should not be punished. As followers of Darwin, we are mandated to encourage them.

Therefore, in the spirit of encouragement, I can, without any reservation, urge you to believe the evolutionary position that the male human dominates the reproductive cycle of the human species. The male human shall breed at will until another male presents himself as a challenger. Then, let the strongest survive and the weakest die as they fight for the possession of the female.

The female human is subjected to the will of the victorious male, assuming she is incapable of resistance. However, the female human can resist mating if her strength allows. Such resistance would lead to a stronger man breeding that woman; a man of such strength that she could not resist. Such mating of stronger individuals would result in stronger children, thus improving the species through natural selection.

In this, we see the cycle of the Darwinian process of human breeding and reproduction and how it benefits the human race by producing stronger individuals and lessening the reproduction of weaker individuals. In this breeding cycle, we witness the pure mechanism of evolution—Survival of the Fittest.

By embracing the Sub-Law of Reproduction we insure that our species will become stronger with each generation and that the process of evolution will continue to select us for existence and not extinction.

ENDNOTES

1   Darwin, Charles, *From So Simple a Beginning*, New York: W.W. Norton & Company, Inc., 2006, pg. 506.

2   Ibid., pg. 489.

3   Ibid., pg. 483.

4   http://whyfiles.org/095evolution/index.html, courtesy of University of Wisconsin Board of Regents.

5   Darwin, Charles, *From So Simple a Beginning*, New York: W.W. Norton & Company, Inc., 2006, pg. 506.

CHAPTER EIGHT

THE SUB LAW OF PLANETARY FAILURE

From the cellar of Dr. Iman Oxidant

TO MY WALKING SACS of bone and water:

A myriad of accidental events led to the formation of our planet. There is no way to know how many billions of years were necessary before the right amount of cosmic matter merged, melted, and exploded into the creation of Earth—a planet that is now capable of not only sustaining life, but producing it. Part of that production was the development of humanity.

It is beyond our comprehension to truly grasp how it all came to be, but we are confident it was not a result of an orchestrated effort at the hands of a creator. With that in mind, we are forced to acknowledge that without a creator, we have no security in this creation. If you think we are secure in our existence, you are betting against the house.

On that "all in" gamble, let us extend the journey into our evolutionary lives with the Sub-Law of Planetary Failure.

It has been my experience that many people truly do not understand the meanings of the words they use. Call it laziness, lack of scholarship, or what not, but let us not fall into the trap today. Here are a couple of key definitions from the New Oxford American dictionary that will serve us well:[1]

Random: *Adjective*—made, done, happening, or chosen without method or conscious decision.

Chaos: *Noun*—complete disorder and confusion; behavior so unpredictable as to appear random, owing to great sensitivity to small changes in condition.

Humans are simply kneaded organic dough and the Earth is the oven that made us rise. We are formed by sheer chance, which is the equivalent of being formed by chaos; chaotic events out of our control and completely void of design and purpose formed this planet and the human species. The probabilities of that happening clearly break every rule of statistics regarding what is reasonable and what is not.

However, as evolutionists we don't analyze those numbers and conclude they give proof of a creator. We decipher those statistics and conclude that they are peculiar, and therefore, deemed inconsequential. An important part of evolutionary science is determining what is worth considering and what is not.

You should be thanking your lucky cosmic dust that scientists have this rigorous scientific method. For without it, we would be stuck believing that our intricately complex planet was *designed,* rather than formed out of ignorant chance. To believe in a creator of this planet is a waste of my highly evolved brain cells. Really, what are the odds?

With human reason as our standard, we have concluded that all things came into existence through the amazing mechanism of blind luck. I must say that only the greatest minds in the world, in the finest educational institutions, could have arrived at this brilliant conclusion about the formation of life: *one day, it just happened.*

That means our elevated intelligence was born from something that had no intelligence. It is mere chance that directed the events that created our life, and chance, by its nature, has no intellect of its own. And when intelligence is absent, logically, ignorance is all that remains.

So in a very real sense, our intelligence, which is so highly exalted, was born from ignorance (chance). As smart as we have become as evolutionists, we know that our foundational beginnings, the origins that we claim, begin with total and utter ignorance.

Indeed, we see the great wisdom and clarity of mankind when we remove God from the equation. Please allow me to break this down into layman's terms as we digest how these facts impact us.

Our planet, at absolutely any time, could implode, explode, or be annihilated by some source that is completely out of our control. This is possible because the converse is also true: our planet was created by something completely out of control. If our origins are from ignorant, chaotic chance, then everything in the past, present, and future will continue to be chaotic chance. Lack of control is the only constant.

Due to all this uncertainty, our limited planetary resources have been a well-publicized issue for the better part of the past century. We know that the human species, due to our unbridled consumption, is putting too much pressure on this fragile planet. Therefore, our panic and anxiety regarding grave threats like global warming and nuclear war are completely justified.

However, the greatest threat to the planet is the source of all these residual threats. This threat is the growing number of mouths to feed, the hands that pollute, and the lungs that exhale poisonous carbon dioxide.

> *Although some species may be now increasing, more or less rapidly, in numbers, all cannot do so, for the world would not hold them.*[2]
>
> — CHARLES DARWIN

We see in the first part of the quote above that world species increase in numbers relatively rapidly. The question then becomes how was Darwin able to determine if our reproductive rates were rapid? What measuring stick allowed him to make that assessment?

The answer comes in the last part of Darwin's statement. Mother Nature determines what is rapid and what is not based on what her resources can sustain and how quickly the species of the planet consume all of the resources necessary to sustain life.

Darwin concluded that the limited resources of the world would be overwhelmed by the population because of the reproductive rates of the species of this planet. Knowing that nature is far superior to man, a checks and balances system was instituted (if that is even the correct terminology because it denotes an intentional thought process) to keep this population explosion under control.

> *Hence, as more individuals are produced than can possibly survive, there must in every case be a struggle for existence, either one individual with another of the same species, or with the individuals of distinct species, or with the physical conditions of life.*[3]
>
> — CHARLES DARWIN

What a fantastic observation! Think this through with me for a second, or better yet, allow me to think it through for you.

The world has always produced more individuals than can survive. Out of necessity then, one characteristic of nature is that strong individuals kill weaker ones. This we already know.

This mechanism of evolution, this struggle for existence is absolutely critical. You cannot have the cake without the batter. A car cannot roll out of a manufacturing plant without an assembly line first putting it together. It is impossible to have the result without the process. We cannot, in any way, have evolution without natural selection. If natural selection does not happen, if this struggle for existence does not occur, then neither does evolution. And if that is the case, then we have no idea where we came from.

This is why the human condition and the adoption of religion are so vexing and alarming to me. As a society, we do not operate under pure Darwinian law when we apply morality to our lives. This insinuates that natural selection cannot occur within humans and it prompts us to ask the questions: What happens when there ceases to be a struggle for existence? What happens when a species does not abide by the laws and iron fist of natural selection/survival of the fittest? What happens when we allow the sick to be healed, the weak to find rest, and the helpless to be born?

The answer is inefficiency—the enemy of evolution! Our very lives become a running contradiction to evolution through natural selection! And that is very dangerous on multiple fronts.

The whole point of evolution is to weed out useless excess so that the most perfected individual breeds and continues its line. Yet, many people prolong the life of the weak by feeding the poor, helping the needy, and healing the sick. We allow marriage to restrict the breeding rights of the powerful. The poorest of all people are producing the most offspring, and the most affluent are having the fewest. According to evolution, this trend should be reversed. Nothing is making sense.

Therefore, our only conclusion is that these inconsistencies are leading us to a population explosion that will consume the resources needed for life and will eventually lead to our complete extinction at the hands of Planetary Failure.

Our current course cannot be sustained if we continue to reproduce without the simultaneous culling of the weaker parts of the population. The facts demonstrate that too many of us are living.

Please understand, my men of intellect, that evolution is not wrong. Our whole scientific belief system stems from the fundamental teaching of natural selection as outlined and explained by Charles Darwin. Therefore, if we do not want to destroy our planet because of a population explosion, we must obey the core principles of evolutionary logic and restrict our population growth by any means necessary. We are currently exercising some caution through the syringe of abortion and the gas of euthanasia, but we must do much better.

If we do not, we can deduce that disaster is coming. If we disregard those instincts that the natural process of evolution gave us, then the scariest environment imaginable will be upon us. We would suffer a loss of life on a scale of multiple world wars. And I for one, will do all that I can to make sure one of those lives is not mine!

THE EXTINCTION OF EVOLUTION

Humans are even allowing this compassion for each other to spill over to other species. We save the whales, adopt pets, and even have entire organizations built to fight for the rights of animals in human courts! That is a bizarre and completely unexplainable outcome of our evolution! Natural selection should have created a selfish human—not a compassionate one.

Darwin noticed the awful messianic habit humans have in trying to preserve life that is not ours to preserve. Humans act like we have stewardship, or dominion, over less-evolved life. That ignorant, primitive mentality must again be blamed on Christianity:

> *Nevertheless so profound is our ignorance, and so high our presumption, that we marvel when we hear of the extinction of an organic being; and as we do not see the cause, we invoke cataclysms to desolate the world, or invent laws on the duration of the forms of life!*[4]

In his remarks, Darwin showed complete outrage. He openly criticized the way humans intervene in nature instead of just acting within it. He called humans ignorant and presumptuous and claimed that our actions are cataclysmic as we try to prevent the deaths of organisms by invoking laws that legislate the duration of life.

Darwin understood that humans, being the dominant species, should end the lives of the weak, not waste resources trying to prolong them! It is not our place to try to save the lives of those that natural selection has chosen to eliminate.

Furthermore, because we are a product of natural selection, this instinct we have to sustain life presents a conundrum that perhaps only the greatest minds in the world can understand. Since we are the most highly evolved, we should be the very first to recognize and eliminate weakened life, yet many of our species are doing the exact opposite...rushing to save it!

The fact of the matter is that the weak are the weak; there is no line drawn to protect one animal from another. It pains me to think that even in our prized universities, the majority of professors act as if Darwin's laws are not applicable to the human species. Oh, what arrogant outrage to pretend we are outside the circle that actually created us! Do you not see that artificially protecting one life just deprives resources from another?

Darwin was scolding us when he observed that we "invent laws on the duration of the forms of life!" Darwin taught that nature should run its course. If a species is going extinct, let it die; another will rise. If a human is dying, let it die. Death is the result of being weak and being unable to continue the biological functions of life. As we rush to save those weaklings, we endanger the entire species due to our limited resources and impending population explosion.

> *As many more individuals of each species are born than can possibly survive; and as, consequently, there is a frequently recurring struggle for existence.*[5]
> — CHARLES DARWIN

Too many individuals are born on this planet. Those fit to survive exist and reproduce. The weak are naturally selected for death—or they should be. As every science professor and public school teacher should tell you, this is nature and the basis of all the life sciences. Failure to abide by these laws leads to the evaporation of resources and eventual Planetary Failure. Humans are the only species not abiding by these Darwinian practices in our daily lives. We are in the heart of a species-wide epidemic and compassion is the virus.

Our actions, as the dominant species, should be in harmony with evolutionary practice, not fighting against it. Thus, it is mandatory that this recurring struggle for existence becomes the mantle upon which humans reside. The strongest must thin out the weakest. If this does not happen, the human population alone would destroy the planet by consuming every last bit of space and every vital resource.

> *There is no exception to the rule that every organic being naturally increases at so high a rate, that if not*

destroyed, the earth would soon be covered by the progeny of a single pair. Even slow-breeding man has doubled in twenty-five years, and at this rate, in a few thousand years, there would literally not be standing room for his progeny.[6]

— CHARLES DARWIN

My fellow evolutionists—those who have heard the call of science, reason, and intellect—please hear my voice. We have limited resources on this planet. If there is not a culling of the herd, and a reduction in the massive number of births among the primitive and poor, we will see all available resources completely consumed. How else can you justify Darwin himself saying, "There would literally not be standing room for his progeny"? The key word is *literally*. He was not trying to be poetic or dramatic; he was being scientific.

Natural selection has attempted to thin out the human race, but again, our sense of morality, stemming from Christianity, has clearly clogged our intellectual pipes. We have this sick idea that we should preserve ailing life. But this will lead to our own destruction, death, and eventual extinction if we do not let our instincts take the wheel. If we do not do this voluntarily, natural selection will force itself upon us and take us all out. It will eliminate the entire human race just like it eliminated the dinosaur.

If my words sound harsh, you are still not grasping real evolution. Letting those who are sick, weak, unhealthy, or undesirable die—whether human, canine, feline, or even a primate—should not make us feel bad or make us struggle with so-called moral hurdles.

When we reflect on this struggle, we may console ourselves with the full belief, that the war of nature is not incessant, that no fear is felt, that death is generally prompt, and that the vigorous, the healthy, and the happy survive and multiply.[7]

— CHARLES DARWIN

We can find it natural and even humane to kill, not evil as the zealots would have you believe. The healthy and the happy should eliminate the sickened and the sad, so is the logic of Darwin. In fact, we should console ourselves; because as Darwin said, no fear is felt when we ease a fellow organism into death.

The basic laws of natural selection grant us the authority to kill. And as we learned, Darwin's esteemed logic cannot be questioned. So please, if you must hang on to compassion, show compassion for the resources of the strong as you accept your place among us by assisting the weak in their exit from this environment.

The frightening issue with present-day humanity is that we live in contradiction to the foundational logic of Charles Darwin. More people are being born today than are dying; we are not answering the call of nature to thin the herd. This is a horrible trend, because it is in direct conflict with evolutionary thought. We can only conclude, because scientists have correctly stated that evolution is fact, that the human species is in line to suffer huge losses for this egregiously unbalanced ratio of births and deaths.

Do you realize we are actually building hospitals that have the sole purpose of finding ways to prolong the life of the weak? We need only look at the Terry Schiavo court case to see that humans will try anything to keep the very weakest of humans alive. Thank goodness the U.S. government finally allowed her death to take place. It is good to see evolutionary ideals lead to that kind of progress.

In this victory we can take heart. We are making progress in the battle toward the acceptance of euthanasia. When we can terminate life at the beginning through abortion and terminate it later for the old, disabled, unproductive, and unwanted through euthanasia, we are starting to take control of the future, at least those of us that have a future.

The facts have been stated clearly, and against them no rational argument can be made. In the near future our planet will no longer be capable of sustaining human life as we know it.

Our unbridled consumption of resources will bring about cataclysmic events that will threaten the human species and the entire planet. Our reproductive rates are too great and our sheer numbers will overrun the planet. Couple this problem with the current lifestyle of compassion, in which we refuse to cull the herd and rid ourselves of the weakened individuals, and we are left with a swarming infestation of humans.

Therefore, humans are the greatest cause of Planetary Failure. The calculated reduction of specific human populations is of paramount importance in order to sustain the planet and the strongest of the species.

ENDNOTES

1   New Oxford American Dictionary.

2   Darwin, Charles, *From So Simple a Beginning*, New York: W.W. Norton & Company, Inc., 2006, pg. 490.

3   Ibid., pg. 490.

4   Ibid., pg. 496.

5   Ibid., pg. 451.

6   Ibid., pg. 490.

7   Ibid., pg. 500.

CHAPTER NINE

THE FRUIT OF EVOLUTION

From the Cellar of Dr. Iman Oxidant

TO MY BUDDING evolutionists:

We have concluded the four Sub-Laws of Evolution. We crafted them from rock solid, indisputable Darwinian logic. We neither exaggerated nor embellished. If anyone disagrees with us on the points we made, they would be disagreeing with Darwin himself—the equivalent of scientific blasphemy.

> *A century and a half of research on all fronts of biology has shown that Darwin was correct in his reasoning natural selection is the general driving force of evolution.*[1]
>
> — E.O. WILSON

Darwin's work is the foundation of modern science. Now we will build upon it by adding another brick.

> *What applies to one animal will apply throughout all time to all animals—that is, if they vary—for otherwise natural selection can do nothing.*[2]
>
> — CHARLES DARWIN

What is true for the hippo is true for the human. We are all under the same system of natural selection. If one species breaks stride with the rest of the species of the planet, the whole machine breaks down and natural selection can do nothing.

We have reached the crux of the matter: if mankind were created equal, then natural selection could not choose between us and evolution would be rendered void. If all members of the human species have an equal right to life, whether white, black, Asian, Hispanic, Arab, or Jew; if all have the right to live without the fear of being killed by an encroaching, more highly evolved party, then all of evolution is nothing more than an outdated thought.

This is the point in history at which we now find ourselves. Do all of us have the right to the pursuit of happiness and the basic right to life, or is evolution true? It is impossible for both of these all-inclusive ideologies to exist simultaneously, for they are in direct competition with each other.

My friends of reason, do you see the massive predicament we have gotten ourselves into?

Tolerance and compassion, combined with the lack of sexual aggression, are human inclinations that are literally destroying us. We have abandoned the instincts that would have us exterminate the sick, steal from the weak, and kill the less-evolved. If we do not accept Darwin's admonishment and let survival of the fittest rule the street, then natural selection can do nothing, and the true creation story of mankind is dead. Not only is evolution dead, it never existed in the first place; for in the absence of natural selection, evolution loses its engine and ceases to have power.

The facts continue to daunt us. The human species, through simple observational study, shares very few common habits with our brethren in the rest of the animal kingdom. Show me sexual restraint; show me the desire to help the needy and cure the sick among the animals of the wild. These things are unique to humans as are many other attributes.

But we are not unique! We are just lumps of organic matter that evolved from the same primordial ooze the rest of the animals, plants, and parasites came from. Yet as we disobey nature by not acting within our animalistic instinct, we risk rendering the law of evolution obsolete.

Logic would contend that because we grew from Darwin's Tree of Life we should be producing evolution's fruit. But instead, we are bearing no fruit of that lifestyle. The ways of evolution are noticeably absent—while morality and religion are painfully prevalent.

Even though we don't widely spread the fact that morality is an anomaly in our world of evolution, the evolutionary elite know this is the thorn in our side. P. Z. Myers from *SEED Magazine* wrote in 2006 that it's "not immediately obvious how a Darwinian regime would foster kindness and charity."[3] He actually goes on to state the same point I have been laboring over, and that is "self-sacrifice for the benefit of unrelated individuals ought to be selected *against.*"[3]

Myers means that displays of self-sacrifice, in the world of evolution, are a detriment and natural selection should act to eliminate them. However, as we know, the human, the most evolved species on the planet, is infested with self-sacrificial habits. We even give tax breaks to charities! We not only allow compassion and self-sacrifice, we reward it!

Simple reason would confirm that if we are the most favored species of natural selection, evidenced by the fact that we are the most dominant species currently on the planet, then we should most clearly mirror the ways of evolution through natural selection. This is our most vexing dilemma, because we are, in fact, the polar opposite of what we should be!

Therefore, we are faced with an almost impossible riddle and are forced to conclude that we cannot learn about humans by studying humans. To keep evolution alive, human habit must be discarded altogether. If we put human habit into the evolutionary equation, it completely changes the answer. Our tendency toward compassion, and our total lack of true animalistic aggression, throws everything out of whack because evolution is supposed to be a fact—yet, we are treading upon it showing it not the least bit of respect.

In our quagmire, there is only one thing left to do: embrace the ways of evolution. It is time to walk within our instinct. We must take what is ours and let the strong press forward and make the weakest recede. That is the only way to protect our future.

We can invoke the analogy of Darwin's Tree of Life to help us do that. Please imagine an apple tree in your mind. It does not bear any pears or oranges, because an apple tree only produces apples. One does not grow an apple tree in hopes of picking oranges—one grows an apple tree to harvest apples. Likewise, when we fully adopt Darwin's Tree of Life, we will start to bear its fruit and our actions will be born from evolutionary thought.

In the seminars that follow, we will be discussing the fruit of evolution by applying the Sub-Laws of Evolution to modern day questions, problems, and phenomenon. We will explain homosexuality, population explosion, proper breeding etiquette, and more through the grid of evolution and Darwinian thought. As we implement evolution into our lives we will see how evolution can finally embrace us and graft us back into its tree.

So behold the vast garden Darwin has planted. Everything in it is good to eat. Let us now harvest its choice fruit and eat the words of evolution.

ENDNOTES

1   Darwin, Charles, *From So Simple a Beginning*, New York: W.W. Norton & Company, Inc., 2006, pg. 439.

2   Ibid., pg. 522.

3   P.Z. Myers, "Bad Religion," *SEED Magazine*, November 2006, pg. 89.

CHAPTER TEN

A FRUIT OF EVOLUTION: HUMAN REDUCTION

From the Cellar of Dr. Iman Oxidant

MY HUMAN plagues:

It is time to preemptively strike at the coming population explosion, something which has been of serious concern for some time. I don't have to tell you that if something is not done soon our planet will buckle under the pressure of the human species, snap like a dried up vine, and leak like a busted pipe. Fortunately, evolution provides a clear direction to follow, but we need to have the stomach to do what is right for the species and for the planet.

Many of the educated have seen the documentary, *An Inconvenient Truth*. The project's website (www.climatecrisis. net) states that the movie, which features former Vice President Al Gore, depicts "one man's fervent crusade to halt global warming's deadly progress in its tracks."[1]

Seemingly, it appears that Gore understands that it is through his leadership that we can save the planet— not the imaginary leadership of some outdated god. Good form, Al.

In this film, Al Gore identifies for us several factors that are causing the destruction of our planet. The very first one requires our deepest attention.

We are witnessing a collision between our civilization and the Earth. And there are three factors that are causing this collision, and the first is population. When my generation, the baby boom generation, was born after World War II, the [world] population had just crossed the two billion mark. Now, I'm in my 50's, and it's already gone to almost six and a half billion. And if I reach the demographic expectation for the baby boomers, it'll go over nine billion.

So if it takes 10,000 generations to reach two billion and then in one human lifetime, ours, it goes from two billion to nine billion, something profoundly different is going on right now.

We're putting more pressure on the Earth. Most of it's in the poorer nations of the world. This puts pressure on food demand. It puts pressure on water demand. It puts pressure on vulnerable natural resources, and this pressure is one of the reasons why we have seen all the devastation of the forest, not only tropical, but elsewhere.[2]

— AL GORE

As I watched Al Gore make this impassioned speech, I could hardly sit still for excitement was racing in my veins. Mr. Gore made a succinct, perfectly executed argument on our bloating population. The only thing left for him to do was finish what he started—strike the gavel and announce his judgment. But he never did. He inexplicably stopped and never offered a solution to the population crisis he laid out!

A few weeks went by and I often pondered why he was not able to give his conclusion. Did he not know the answer? Was he confused by his own beliefs?

One morning, while I was sipping English tea in my favorite Boston cafe, it hit me. Mr. Gore is a politician and he always will be. He is driven by the latest polls and the moods of the masses. I, however, am not. I have no need for popularity. Natural selection is the only approval I seek.

Therefore, I will not stop where the political ice gets thin. I will articulate the truth that was apparently too inconvenient for Mr. Gore to speak openly. This truth is that the population of humans is running rampant and is out of control. A reduction is mandatory. This is the prevailing thought of scientists who embrace evolution. The need to reduce the human population is certainly fruit that falls from Darwin's Tree of Life—or death—however you want to look at it.

> *Even slow-breeding man has doubled in twenty-five years, and at this rate, in a few thousand years, there would literally not be standing room for his progeny.*[3]
>
> — CHARLES DARWIN

> *Although some species may be now increasing, more or less rapidly, in numbers, all cannot do so, for the world would not hold them.*[4]
>
> — CHARLES DARWIN

Darwin's thoughts on human population levels have penetrated deep into intellectual communities. The enlightened understand that humans are becoming a plague to this planet, and if something or someone does not intervene we will all suffer greatly.

Prince Philip, the grandfather of Prince William, the future King of England, no doubt understands that mankind is a plague upon the Earth and becoming a serious threat. He said, "If I were reincarnated I would wish to be returned to Earth as a killer virus to lower human population levels."[5] I can only hope he is training his grandson in this splendid philosophy.

Prince Philip is scarcely alone in his views. Many prominent evolutionists and scientific leaders also take heed from Darwin's warnings. One such man is Eric Pianka, professor of zoology at the University of Texas at Austin.

Pianka penned the textbook, *Evolutionary Ecology,* and more than one hundred scientific papers. He was a 1978 Guggenheim Fellow, a 1981 American Association for the Advancement of Science Fellow, and a 1990 Fulbright Senior Research Scholar. Pianka is recognized as an expert by the brightest scientific minds and his influence is deeply felt in scientific circles.

In addition, I have heard he even has a photograph of himself dressed up like Charles Darwin hanging on his office wall.[6] That makes him an honorary member of The Institute of Progressive Lineage.

Let us see what this honorable and often-awarded professor has to say regarding our current subject matter. In Pianka's article, "What nobody wants to hear, but everyone needs to know" he says:

> *I am convinced that the world WOULD clearly be much better off without so many of us.*[7]

Dr. Pianka! But whatever do you mean?

> *I actually think the world will be much better off when only 10 or 20 percent of us are left.*[8]

Thank you for the clarification, Dr. Pianka.

With the population of the world now at approximately 6.5 billion, Pianka's reduction figure of 90% would mean that approximately 5.85 billion people would need to die before the world would be better off.

How 'bout them apples? That is the kind of fruit evolution leaves behind. When a highly educated professor like Eric Pianka allows pure Darwinian logic to rule his thoughts, one can likewise begin to imagine some great solutions to the world's problems.

Now be careful not to sugarcoat these statements. Pianka did not say the world would be better off if only 10 or 20 percent of us lived. He said the world would be better off when only 10 or 20 percent of us *are left.*

This is obviously a man who understands evolution and natural selection. In fact, his train of thought carries the same element of clarity Darwin had when he mentioned that some races of humans are probably not human at all.

More of Pianka's writings confirm that if we humans do not take this population problem into our own hands, natural selection will do it for us:

> *We need to make a transition to a sustainable world. If we don't, nature is going to do it for us in ways of her own choosing.*[9]

If we subscribe to evolution, this is the only outcome we can foresee. If we don't aggressively address our problems, then nature will do it for us—and that is not good for anyone.

Again, Eric Pianka agrees with our Sub-Laws that humans are brazenly stepping out of sync with the rest of the animal kingdom:

> [The human species] *thinks it can violate the rules of the natural world and that it can grow indefinitely— humans think they can defy the rules. We are Homo the sap, not sapiens (stupid, not smart).*[10]

Personally, this is one of the great mysteries of life. We once again see the reoccurring, fundamental problem with humans: we are breaking all the rules. This is bizarre behavior for the most advanced and dominant species. If a law does not have to be adhered to, is it really a law? A law is validated only when it is backed by true authority.

All of this disobedience to the laws of nature can only mean one thing—we will soon see the authority of natural selection manifested in death to the human species.

According to the writings of Darwin and Pianka, humans are defying nature. That cannot be allowed. If we don't change, natural selection may force extinction on the entire race. It is necessary for the few of us, who are intellectually enlightened and depend on our reason to guide us, to press upon and exterminate the others. If we don't voluntarily trim the herd, it is possible the entire human species will be pushed into extinction.

One obvious way to eliminate the plague of our bloated population is through killer epidemics. Some might point to the AIDS epidemic in Africa as being something that will siphon our population. Yes, it will do that. But, according to Pianka, even AIDS has its limits in helping us reduce our planetary population levels:

HIV is a pandemic spread worldwide. It's increasing in frequency in a lot of places and it's a big concern to everybody. But, it's not going to be the one that gets us because HIV acts too slowly, it lets us live several years while it passes itself on to new hosts. HIV is no good, it is too slow to control human populations.[11]

HIV is not the right epidemic to wipe out enough humans! I was really holding out for that one, but there may still be hope:

Now let's consider some other viruses — Ebola zaire has potential. It kills nine out of ten humans. It's never gotten out of Africa because it is so virulent it kills everybody before they can move. I mean it kills you within a day or two. You can only catch Ebola zaire by direct contact with a human who's infected. It causes you to bleed. It breaks capillaries and you bleed out your orifices and if you touch somebody who's sick with it you get it and you die, too— nine times out of ten.[12]

— Eric Pianka

As you can see, Ebola-Zaire may have potential. We should prepare for the worst:

Humans have been very lucky not to have experienced a worldwide plague for a long time, one is overdue.[13]

— Eric Pianka

Darwin would be so proud!

As anyone can rationalize, our evolutionary reality confirms there are two things that must be done about the human population, and they need to be done now:

1. Restrict the number of births
2. Reduce the existing human population

Now we are seeing the sweet Fruit of Evolution.

To give some direction to our efforts, I will take the liberty of suggesting some lifestyle changes the human race must adopt in order for us to live an authentic evolutionary life—a life that is in stride with the foundations of natural selection.

I think it is obvious that the most important strategy we can use to control our population is abortion. I understand that society has matured to such a level that abortions are now no more than a mere afterthought. Going a step further, however, I believe abortion can be more than just individual family planning—it can be species planning.

The apparent mandate of species planning allows us to conceive guidelines for parenting, and even grant or deny Parent Permits. Abortion can be our governor of such when certain permits are violated. Wearing our evolutionary thinking caps, we fully recognize that as more and more of the lower races are allowed to breed, more and more resources are wasted on keeping these less-evolved accidents alive.

Furthermore, with each inferior human born, the average evolutionary status and intelligence of the species diminishes slightly. That creates an evolutionary regression, which causes the species to move backward.

I propose conducting an evolutionary analysis that will help us identify those classes of people who should be denied Parent Permits, and be legally persuaded into having an abortion when their violation demands it. This represents the very first fruits of our evolutionary labor and in no way should this limit or restrict what we need to do in the future. But, as a starting point, the six variables we are now going to list are a healthy first crop. These variables would all need to be considered before a Parent Permit is issued to the mother.

Those of you who may have logistical concerns on how to facilitate this program need not worry. An international socialized health care system can easily control this scenario. As major governments of the world remove health care from the private sector and institute national health care systems, it becomes very foreseeable that an international governing party (like the United Nations) could unify those systems into what I have named,

"Global Care," the international socialized health care system of the future.

On that progressive thought, let's create our six-point checklist for Parent Permits.

1. Financial Stability: The parent must demonstrate financial means so that the burden of feeding, clothing, and educating the child does not fall on others. This reasoning can have no enemy against it. Even Al Gore stated in his acclaimed documentary that the "poorer" nations are the ones to blame for our population overload.

2. Social Status: Those in the upper classes are better looking than those in the middle classes, and by default, much better looking than those in the lower classes. With that in mind, prospective parents must demonstrate that they are members of the social elite and possess the sexually desirable characteristics present in those with large bank accounts. If we are going to improve the population with each generation, we must be unwavering on this point. As Darwin instructed, we cannot be careless enough to allow the worst to breed.

3. Favored Race: The lower races are just that—lower. To improve our evolutionary status, any person who is not a member of the most evolved race should be sterilized and not be allowed to give birth to a child. Such a child would automatically be less-evolved than their counterparts in the higher races. To identify the favored race, we will rely on Darwin's research and reference to the race of Newton and Shakespeare, which is the white European strand of humans and subsequent descendents of that race. Those of mixed blood will also be sterilized to prevent further contamination.

4. Blood Testing: After conception, embryos will need to undergo blood work to determine if any genetic ailment is likely; obesity, diabetes, arthritis, mental illness, etc. It is of vital importance that any condition detrimental to the human species be eradicated from the face of the planet before it is allowed to contaminate it.

5. IQ Testing: All prospective parents need to be tested for intelligence. Based upon a predetermined statistical average, we should allow only the brightest among us to reproduce. We need to be certain that each child born will help us move toward a higher standard and improve upon our evolutionary development.

6. Faith Deficiency: Adherence to religion is a deficiency in the human species. In order to maximize our productivity, any tendency noted on the part of the parents to lean toward the fallacy of religion must result in a complete termination of their genetic line. In this case, the abortion of the fetus will be carried out through the euthanizing of the parents. Religion is a disease—a disease of the mind—and cannot be tolerated.

In his book, *The God Delusion*, author Richard Dawkins sets up his platform brilliantly just by the very title. The author could have called the belief in God a fallacy, a mistake, a mishap, but instead he called it a "delusion."

The New Oxford American Dictionary defines "delusion" this way:

> *An idiosyncratic belief or impression that is firmly maintained despite being contradicted by what is generally accepted as reality or rational argument, typically a symptom of mental disorder: the delusion of being watched.*

The phrase of interest here is "a symptom of mental disorder." Dawkins realized that believing in God is delusional. Thus, faith in God in and of itself is a symptom of a mental disorder. Does that not bring a brilliant light to the subject?

By definition then, faith in God is something that can and should be treated medically like all other mental disorders. Just like we should eradicate schizophrenia, we should eradicate the sick souls who hold on to a delusional belief in God. We cannot allow these disorders to continue in the human population.

Belief in God is considered to be a mental disorder that is hereditary and passed down by individuals within certain communities and races of people. Therefore, our conclusion, based on evolution's insatiable appetite for improving the human stock, can only be one thing: faith in God, on the part of the parent, should lead to the complete genetic line termination.

On a quick side note, I have always had this question: Why would any organism that is created by natural selection, an inherently selfish mechanism, ever create an entity to worship, pay homage to, and serve other than themselves? It is precisely the opposite reaction one would expect from the most advanced product of natural selection. This is just another way we are breaking the rules, I suppose.

What we have listed above is a good start in evaluating who should receive a Parent Permit and who should not. However, controlling future births does not fix our current population problems. One might even argue that adjusting the current population might be the best way to start because you can't have children if you are already dead.

Enter the Euthanasia Initiative: Aborting a fetus before it becomes a nuisance is a good step. However, there is still a huge problem with the existing population. According to evolutionary scientists, the number of people living on the planet is already too high. Because no credible experts have stood against him, we must conclude that Eric Pianka was correct and that his recommendation of a 90 percent reduction in population is ideal for a planet of this size.

I understand that some may think that being humane is important. But remember that Darwin said, "no fear is felt" in death and that it is "generally prompt." Considering those words, it does not seem relevant whether the living really believe something is humane or not. We just need to be obedient to our instincts.

Saying that, I recognize Darwin stated that truth about the emotions of death before he actually experienced it, but we should not allow that to discredit his findings. We should always remember that Darwin based most of his theories on things he never experienced or witnessed, so there is no reason to waiver on this one if we didn't waiver on the rest.

To ensure that the weaker races and the genetically compromised pass away easily, we must engineer mechanisms that eliminate them efficiently. This would be easier on both the living and the dying. I am sure if we look into our rich history, we should be able to find examples of how to do this effectively.

There is no limit to what mankind can accomplish when we put our heads together and stick strictly to evolutionary logic. In Europe we already find progressive societies that understand the importance of a timely death. In Switzerland, it is now legal to prescribe euthanasia when a person suffers from a mental disease.

Ultimately, Switzerland's legislation allowing these aggressive measures will not just be an isolated incident, but rather the first domino that falls. Truth be told, if certain people cannot be cured from delusional religious beliefs, I think this medical procedure of euthanasia will prove to be the most effective treatment for their condition, whether in Switzerland or anywhere else in the world.

Furthermore, there is evidence that when someone is put to sleep, it is a wonderful thing; the doctor who facilitates the death of such a "patient" feels good about what he has accomplished:

> *I administered first a large dose of sleeping pills and she drifted into a deep sleep. After five minutes, I gave her a muscle relaxant which lames the breathing apparatus and the heart—this is the way we do it still. Her breathing stopped, her heart stopped, and she died quietly a few minutes later. It was an utterly criminal act then. But I did not feel that I had committed a crime—I had done something good for somebody.*[14]
>
> —Dr. Rob Jonquière,
> "What it's like to help someone die," BBC News

Interestingly, the "patient" Dr. Jonquiere was treating *wanted* to die, there is no doubt there. She wanted to die to escape suffering she was not yet enduring; she was afraid that one day she would experience too much pain. I suppose medical practitioners were predicting an awful future. Therefore, instead of dreading the day when too much pain would come to her, she allowed her doctor to put her down.

The patient said the following to Dr. Rob Jonquière: "The fear of unbearable suffering was worse than the suffering itself."[15] She wanted to die out of fear; a fear that was not yet reality.

Thankfully, a good doctor, who was willing to kill her, was there to do the right thing. There's no use living in fear; we can't allow the fear of the unknown to worry us too much, because it is unbearable. That is a fine justification for euthanasia.

This heartwarming story illustrates that euthanasia can be a wonderful part of life. But more than just the terminally ill should be a candidate for such a process. The obvious starting point would begin when people reach their inevitable end—but keep dragging on, and on, and on. The decrepit elderly must know they would be helping the entire world by exiting just a tad bit early and saving us the aggravation of tolerating their degenerative states. But then again, people just do not know when enough is enough.

These decaying humans are dangerous behind the wheel, slow to speak, slow to think, and slow to move. They are often messy eaters and smell bad. Yet, most governments of the world go to great pains to ensure these people are able to consume huge quantities of resources while contributing nothing to society. This is lunacy and irresponsible to humanity.

There is no question that within our Darwinian lifestyle these people should be euthanized. The resources they are consuming should be utilized by others more fit to survive in our environment.

In addition to the elderly, we can also cite every reason used to justify abortion to justify euthanasia of the living. The weak are the weak, whether a fetus, a twenty-year-old, or a ninety-year-old.

If we review the Sub-Laws of Evolution, we will see that the strategies of euthanasia and abortion we have discussed do not interfere, conflict, or contradict any premise laid out in those laws. As we stated earlier in these seminars, you can be assured that the strategies I have laid out before you are based on the core beliefs of Darwinian evolution. Therefore, this is the Fruit Darwin's Tree of Life produced: death to anyone not fit to survive guided by the unforgiving, pitiless sword of natural selection.

My dear accidents of nature, before you despair about the amount of work and correction we have to do, I also come bearing great news and glad tidings. Any lord of war would know that one of the best methods for destroying an enemy is to destroy it from within. I offer you a beautiful example of that happening. This was reported on November 12, 2006, in the *UK Times*. It was written by Peter Zimonjec, and inspirationally entitled, "Church Supports Baby Euthanasia."

> *The Church of England has joined one of Britain's royal medical colleges in calling for legal euthanasia of seriously disabled newborn babies.*[16]

In the same article, Tom Butler, Bishop of Southwark, states:

> *It may in some circumstances be right to choose to withhold or withdraw treatment, knowing it will possibly, probably, or even certainly result in death.*[17]

This is a shining moment; we are finally starting to secure the strength of our species. And we are doing so by using the most foundational evolutionary truth: eliminating the weak and allowing the strong to survive. How can we go wrong when we are so perfectly Darwinian?

Furthermore, and maybe just as important, Bishop Butler's actions show us that the foundation of the Church is breaking down. Fear of the unknown is overriding their faith, and rightfully so. Faith is a myth, and a blind, feeble one at that. In this situation, it is being publicly stated, by the Church, that it is best for everyone that the defenseless, voiceless child dies.

I am fully convinced that if this argument can be won, then any argument can be won! After all, it's a bishop doing the arguing— simply brilliant! This is massive progress; not only are we free from sin because of evolution, we have even convinced bishops they are free from sin. That is when you know you have your enemy in a choke hold!

With this monumental move we can rest assured our resources will remain plentiful as we expand the definitions of who should live and who should die. Such evolutionary moves will make our species stronger. I wonder if poor Bishop Butler even realizes his faith is a genetically-based mental disability that will have to be addressed down the road. Yet, he helped pry open the coffin for the defenseless infants being euthanized with the blessing from the church. Talk about getting burned by dancing with the devil! We are clearly seeing the prickly Fruits of Evolution, but it is only a sampling of what is soon to come.

Through abortion, we cull the unwanted herd of humanity before we even have to deal with their first breaths. Through euthanasia, we remove the malformed of nature; those who do not meet the high criteria set by the finest of our species. Based on the reasoning of Darwin, we have provided examples and direction regarding how to navigate the terrain of resources, reproduction, and the reduction of our species.

In light of this progress, I feel compelled to remind all the itching ears listening that with no sin, "Thou shall not kill" has no authority. We understand humans are not created equal, so the strong have every right, and indeed the obligation, to eliminate the weak. By adopting the practice of euthanasia, the human species imposes a self-corrective measure; this way natural selection does not have to eliminate the entire human species because of our catastrophic excesses.

Even with the massive amount of Darwinian logic I have offered, I can sense the slow-witted, morally-shackled man will *still* ask, "Do we have the right to exterminate the less-evolved so callously?" Allow Charles Darwin to answer that directly, so there are no misconceptions:

> *The enquirer would next come to the important point, whether man tends to increase at so rapid a rate, as to lead to the occasional severe struggle for existence ... Do the races or species of men, whichever term may be applied,*

encroach on and replace each other, so that some finally become extinct? We shall see that all these questions, as indeed is obvious in respect to most of them, must be answered in the affirmative, in the same manner as with the lower animals.[18]

Darwin's clarity is appreciated. He concludes that man increases at such a rapid rate that a severe struggle for existence will happen between, as Darwin said "species of men," resulting in one (or many) finally becoming extinct.

In other words, those humans who fail to improve will be eliminated. This *struggle for existence* between humans is unavoidable, inescapable, and an evolutionary certainty. Humans, who are better adapted and more perfectly evolved, will survive and reproduce. Humans who are not as well adapted will be singled out for death.

We have now tasted the fruit and inevitable outcomes of evolution.

ENDNOTES

1   www.climatecrisis.net.

2   Al Gore, *An Inconvenient Truth*, Paramount Pictures.

3   Darwin, Charles, *From So Simple a Beginning*, New York: W.W. Norton & Company, Inc., 2006, pgs. 490-491.

4   Ibid., pg. 490.

5   http://www.freerepublic.com/focus/f-news/1145419/posts.

6   Paul Joseph Watson & Alex Jones, "Dr. Death Gets FBI Visit", www.Prisonplanet.com, April 6, 2006.

7   Pianka, Eric, "What nobody wants to hear, but everyone needs to know," http://uts.cc.utexas.edu/~varanus/everybody.html.

8   Pianka, Eric, "The Vanishing Book of Life," cited from http://www.zo.utexas.edu/courses/bio373/Vanishing.Book.pdf, 33.

9   Pianka, Eric, "What nobody wants to hear, but everyone needs to know," http://uts.cc.utexas.edu/~varanus/everybody.html.

10 Pianka, Eric, "The Vanishing Book of Life," cited from http://www.zo.utexas.edu/courses/bio373/Vanishing.Book.pdf, 23.

11 Ibid., 24-25.

12 Ibid., 25.

13 Ibid., 21.

14 Jonquière, Rob, "What it's like to help someone die," BBC News, November 29, 2001.

15 Ibid.

16 Zimonjec, Peter, "Church Supports Baby Euthanasia," *UK Times*, November 12, 2006.

17 Ibid.

18 Darwin, Charles, *From So Simple a Beginning,* New York: W.W. Norton & Company, Inc., 2006, pg. 784.

CHAPTER ELEVEN

A FRUIT OF EVOLUTION: LOWERING THE BREEDING AGE

From the Cellar of Dr. Iman Oxidant

MY FRUSTRATED amigos:

Ever since you were mere boys you have had certain hormonal urges that have been cruelly suppressed by the overriding Christian culture. There is no doubt that your natural rights as men have been cut short and you have been emasculated. Pathetic traditions have robbed you of your productivity and stolen chances that can never be returned. More sympathetic to this plight, I cannot be. Because you have suffered, humanity has suffered as well.

Show me a lion that offers a reprieve to the lioness, or a baboon that seeks courtship before consummation. Again, humans fail embarrassingly in comparison to the sexual actions of our animal brethren. We must allow nature to bring out the raw sexual aggression in each of us. We cannot allow artificial laws of governments to determine when man can mate, especially when it impedes the drive that nature herself instilled into him.

Some might attempt to refute this line of thought by saying our culture offers suitable guidelines that should be followed in terms of sexual etiquette. But culture is fleeting! It changes with time and fluctuates within the population. Unlike nature, culture is not a binding force—it can be manipulated by political process and religious overtones.

No, culture is not intended to define our reproductive rights. Nature must be our benchmark on such issues. We can look at what nature intends and *then* determine what we ought to do. So, to discover the appropriate age to begin reproduction, we should look at the natural capabilities of the individual.

At this point in human evolution, most boys and girls should be able to breed by the time they reach adolescence in their early teens. Obviously, those ages vary from individual to individual; we have even heard of eight-year-olds capable of reproduction. I would think, however, that over time, the breeding age of humans may drop to even younger ages.

One of the most striking differences between humans and other animals is our abnormally late rate of sexual maturity and subsequently delayed breeding cycle. The tardiness of this maturity is inexplicable when held against the standards of natural selection. This is another evolutionary contradiction and a weakness we must improve.

I realize I have mentioned ages that might be startling to some of you. Let me explain, using evolutionary principles, why humans should start breeding at the earliest possible age their bodies will allow. First and foremost, the notion of marriage, especially abstinence before marriage, has now been thrown under the bus because of the Doctrine of Sin. Therefore, there remains no convincing moral argument to stop humans who are capable of breeding from doing so.

Some might contend that there are some social and/or financial constraints regarding the rearing of children, but as we reduce the overall human population through abortion and euthanasia as we discussed earlier, those suitable for reproduction should have ample resources at their disposal.

It should be clearly understood that I am not encouraging a general population explosion, for I am speaking solely in terms of Darwin's "favored races" in this argument. Through the reduction of the less-evolved, our population will be, once again, in check and under control. So, please use your deductive reasoning skills and understand that we are building upon all previously discussed evolutionary paradigms. With each successful birth of the higher evolved, the human species will show slight improvement.

There will always be however, among the new births, both the strong and the weak. So as we implement our strategies, and repair the damage we have done, we must constantly cull the herd even from the remaining stock that survive the first purging.

According to a report issued in 2000 by the Centers for Disease Control and Prevention, the average age for a woman in the United States to bear her first child is twenty-five. This is evolutionary lunacy! A twenty-five-year-old female of the "favored race" should have kicked out a solid baker's dozen by this time. Nature has provided the capability for that woman to be bred at the age when her body begins menstruating. Modern generations are, therefore, artificially doubling the age when women are first conceiving. Once again, we are breaking the rules of nature, which is compromising the vitality of our species.

If we shift the average age a woman of the "favored race" experiences her first pregnancy to the pre-teen years, then we will more than double the rate of evolutionary improvement and ensure the advancement of our species.

Therefore, this logic holds fast; if nature deems it possible for a woman to be bred at the age of eleven, who are we to go against the powerful hand of nature? For if a woman can be bred, I can guarantee there is a man willing to try—regardless of her age. This is the way of evolution, so it is futile—and dangerous (for the girl)—to fight it.

Men should not subdue their natural reproductive drives—we should act on them. If we do not act on our primal urges, nature will look upon our inaction as weakness. If the male is strong enough to sexually engage the female, then his age is irrelevant. There is no moral standard that prescribes when an individual should become sexually active, or when one should quit. The strongest male should breed the most desirable female(s) that he is able to possess. Age does not matter when the basic urge to reproduce is present.

However, for evolution to produce a more favorable result, older females should be bred less frequently as younger, more appealing females come of age. I don't pretend to know what age the female would become repulsive to most men; I would assume it would differ for each woman. However, a woman would reach that "retirement" milestone more quickly as younger ones become more available to the males. Perhaps the age of thirty would be a good time to turn them out to pasture. For by then, under the system of evolution, these women would already have fifteen or more children. Quite certainly, they would have a body that—if not completely worn out—would be undesirable. So as you can see, living by evolutionary standards changes everything we now know.

In the reproduction relationship we must accept that the female is only a bystander and need not be a willing participant if the male, who has defeated the other males, can also subdue her. The female has no equality with the male, intellectually or physically, and therefore, has no opinion as to whether or not the victorious male should breed her. Women are inferior in every way; Darwin's evolution has made that abundantly clear. The reasoning of the woman is beneath that of the man, and if a man deems it desirable to breed a woman, that woman can present no feasible argument that will have merit or authority. It's like this: boy meets girl, boy takes girl—period.

In the competition to possess women, a man's right to breed will be determined by his set of skills and strengths. It does not matter if he is thirteen or eighty-one; all that matters is that he can defeat his opponents. Evolution will grant him authority if he is the strongest, most capable man available to breed the woman. It may come down to the man possessing what Darwin called "special weapons"—weapons that would both defeat his adversaries and be useful to subdue the female if needed:

> Generally, the most vigorous males, those which are best fitted for their places in nature, will leave most progeny. But in many cases, victory will depend not on general vigour, but on having special weapons, confined to the male sex.[1]
> — CHARLES DARWIN

Keep in mind that defenses are also very important in determining who takes possession of the female. Darwin stated that "the shield may be as important for victory, as the sword or spear."[2]

Taking all of this into account, one can easily conclude that desirable women will spend most of their breeding lives in a state of impregnation until the population of the males willing to breed her completely ceases, or until her body is incapable of bearing children.

As we mentioned earlier, by itself this should not significantly impact planetary failure or population explosion, because we are acting on the assumption that we have increased the practice of abortion and euthanasia on the lesser-evolved human populations and are only expanding the breeding of the "favored races." What impacts planetary failure in a negative sense is when masses of under-evolved women give birth and add to the plague of subhuman children. Therefore, desirable women should be bred in perpetuity, and inferior women should not be bred at all. This is the Fruit of Evolution.

This action would place a huge demand on desirable women. Such demand would foster relentless competition between men, which is the bread and butter of evolution. Natural selection thrives on that struggle. Initially, when undesirable women are taken out of the breeding population, there might be many more men than women. Such an imbalance would push the demand on desirable women to such a level that they would not be able to walk down the street without being pursued.

This situation, however, would recede in time. Eventually, natural selection would reduce the number of men far below the number of women. For like the lion, the gorilla, and the elephant, one male can service many females. As the struggle for existence and the struggle to reproduce become fiercer and more deadly under the evolutionary regime, we will see such a sifting take place.

In conclusion, smart evolutionary logic teaches us that desirable women should live in a state of perpetual impregnation from the initial onset of puberty until they are incapable of reproducing. Likewise, the male shall mate when he becomes capable and whenever he so desires. His age is irrelevant.

I know we seem to have a long way to go before enough people die for us to reach evolutionary equilibrium, but if we take heart and persevere, I know we will make it.

My highly evolved few,

Please take and eat the Fruit of Evolution from Darwin's Tree of Life. May your firstborn come quickly, and may many more follow.

ENDNOTES

1   Darwin, Charles, *From So Simple a Beginning*, New York: W.W. Norton & Company, Inc., 2006, pg. 506.

2   Ibid., pg. 506.

CHAPTER TWELVE

A FRUIT OF EVOLUTION: THE EVOLUTION OF HOMOSEXUALS

From the Cellar of Dr. Iman Oxidant

TO MY STRAIGHT shooters:

Understandably, the homosexual question is one that evolutionists have either ignored or avoided. Up until now, our work has enjoyed the support of the homosexual community. I suppose that is by default, because most religious communities openly reject the homosexual lifestyle. They invoke one of my favorite Christian clichés: "Hate the sin but love the sinner." I still don't know what that means.

Therefore, the homosexual community looks elsewhere for their worldview, and rightfully so. They seek refuge just like everyone else does.

I presume, however, that after the last word of this seminar is spoken, the homosexual population will once again be a broken band of banished refugees. Regardless of our personal opinions on the matter, one thing is certain: the crosshairs of natural selection are locked squarely on them—more so than any other demographic of the human population. Evolution has decided to systematically shoot down the homosexual.

I acknowledge that we risk the alienation of the gay and lesbian community with our evolutionary imperatives, but I do not fear that, for we've already chanced the estrangement of Christians, women, and inferior races. The truth of evolution obligates us to press on in the face of adversity.

We are champions with a cause. Our mantra and motto stand like the Leaning Tower of Pisa—you can't fight science. The inferior, the sick, the religious, the ugly, and the homosexual will not survive the threshing sickle of natural selection when it bears down upon them. That may sound cold, but it is how the matter truly stands.

As we dive into this, we must acknowledge one irritating issue with homosexuality—if it is caused by an actual genetic trait or if it is a learned behavior. In the past, homosexuality has been treated as if it is something only a deviant mind could indulge in. For an example, criminologist Herbert Hendin once remarked:

> *Homosexuality, crime, and drug and alcohol abuse appear to be barometers of social stress. Criminals help produce other criminals, drug abusers other drug abusers, and homosexuals other homosexuals.*[1]

This notion of course is tied to the archaic, religious belief that homosexuality is tied to sin. However, the existence of sin has been sufficiently refuted, and therefore, we must look elsewhere for the cause of homosexuality.

As an atheistic evolutionist, I must look at the natural to explain all tendencies and behaviors. As much as it pains me to say this, our brethren in the genetics fields have failed to produce, or reveal, the specific genetic trait that causes homosexuality. This would cause some to claim that social pressures may be the cause of homosexual behavior and that it is not necessarily a genetically instilled sexual orientation.

I stand here to argue, however, that if there is a learned behavior, then that behavior must have the ability to be learned. For an example: if a runner increases his speed through training and hard work, he can never increase it beyond his genetic ability. The runner will eventually hit a limit to what his body can do, regardless of training technique.

This evolutionary truth also impacts the intellect. If one is a great mathematician, being able to calculate numbers is a learned skill. But that learned skill could not be acquired if the person did not have the intellectual capacity to learn it. The intellectual capacity of the mathematician stems from unique genetic identity. A mathematician can only do what he is capable of doing and cannot go beyond his inherent genetic limitations.

Likewise, if a social environment to "experiment" with homosexual behavior influences a person, that behavior could not sufficiently be adopted and overwrite the powerful instinct of heterosexuality if there were not some leanings in genetic identity already present.

At some point in the homosexual lifestyle, a natural inclination has to emerge for the person to renounce heterosexuality and fully embrace homosexuality. And by personal testimony of many homosexuals, they claim their same-sex lifestyle is, in fact, the natural expression of their true self. This would certainly indicate that homosexuality is a natural inclination that is built into their evolved genetic make up. As you will see, such a transformation in sexual appetite cannot be cavalier because it sets the person up for genetic termination at the hands of natural selection.

> *Generally, the most vigorous males, those which are best fitted for their places in nature, will leave most progeny. But in many cases, victory will depend not on general vigour, but on having special weapons, confined to the male sex.*[2]
>
> — CHARLES DARWIN

The astute observer will see that the lesbian does not quite fit the same mold as the gay man. If a man wants to breed a woman he will breed her. It does not matter if the woman is a lesbian, because as we have learned, mating does not have to be her decision. Therefore, even though a lesbian may try to evade her pursuer vigorously, it is possible that her genetic line may not stop with her.

The gay man is a wholly different story. If truly homosexual, these men will have no desire to have sex with women. Evolution would look upon them as lacking vigor, and deem them completely incapable of continuing their genetic line.

And this leads me to say a few words on what I call Sexual Selection. This depends, not on a struggle for existence, but on a struggle between the males for possession of the females; the result is not death to the unsuccessful competitor, but few or no offspring.[3]

— CHARLES DARWIN

I don't mean to be cheeky here, but Darwin does not say anything about a man's ability to possess another man. The evolution of a species only continues when a male fights another male for the possession of a female. If a male is to recuse himself from that fight, it is impossible for him to carry on his genetic line. The result of such inaction is natural selection removing his entire DNA strand from existence.

One should know that the intent of evolution is to ensure that the genetic line of a population continues. The only possible way DNA can be propagated is when a male and female mate and reproduce. Without this basic function, all human life on this planet would end within a century. If every human—and every other animal that depends on sexual reproduction—became fully homosexual, then the planet would lose the majority of its life forms, and the entire ecosystem would be irrevocably changed.

Due to the inherent complexities and dangers of homosexuality, this group is, without question, targeted for discontinuation by evolution. This is not to say that more homosexuals won't be born, but those who are born would represent the end of a line—a genetic dead end. What else would you call something that natural selection has determined to be unsuitable for reproduction?

The overriding theme of evolution is constant culling—and constant improvement. If a man has no desire to breed women and no desire to father offspring, natural selection will eventually remove his genetic line from the gene pool. By default, therefore, homosexuals are deemed weak according to the tenets of evolution. They will die and leave no progeny, and their fingerprint on this planet will be forever removed and forgotten—almost as if they were never here in the first place.

Homosexuals will simply be branches trimmed from the Tree of Life.

ENDNOTES

1   Horton, Richard, "Is Homosexuality Inherited?", New York Review of Books, July 1995.

2   Darwin, Charles, *From So Simple a Beginning*, New York: W.W. Norton & Company, Inc., 2006, pg. 506.

3   Ibid., pg. 506.

CHAPTER THIRTEEN

A FRUIT OF EVOLUTION: THE RISE OF THE DARWINIAN LEADER

From the bunker below the cellar of Dr. Iman Oxidant

MY FAITHFUL followers:

This will be the last of my seminars on The Sub-Laws of Evolution, and consequently the Fruit of Evolution. The Institute of Progressive Lineage has made it possible for us to have an in-depth review of Darwinian practices, and for that, I am grateful (in the same way in which you are probably grateful to me).

Through our exploration of evolutionary logic, we have uncovered many truths about our founding father, Charles Darwin, and have learned from his very words. Because he was the expert, it is impossible for us to be mistaken about anything we have concluded. The only way for us to be mistaken about our conclusions is if Darwin himself was completely wrong about everything.

As men of science and great reason, we reject the notion that Charles Darwin was wrong. Consequently, we believe that not only are women inferior to men, but there may actually be races of men who are not even human at all. We believe equality is a myth and on that Darwinian rock we shall stand.

Therefore, armed with evolutionary science, we will take the necessary next step. We will force survival of the fittest on a world that exists genetically beneath my enlightened and highly evolved feet. The enquirer may still question if we have the moral ground to do so. The Doctrine of Sin showed us that no real morality exists, because no immovable standards have been set that can be proven to be right and the sole source of justice.

At this time, one could say we have made our points with great efficiency, and nothing else remains to be said. But that would be a mistake. Our discussion would be lacking if we did not discuss one last item.

What would a true Darwinian leader in the modern world be like? Who would embody all of the Sub-Laws with conviction, authority, and purpose?

Yes, indeed, to hear a speech from a perfect Darwinian leader—a speech that whole-heartedly embraced Darwin, his jargon, style, and absolute resolve—would be very enlightening. Let such a speech flow from a man who is strong enough to courageously espouse the ways of evolution; one who could reveal its Fruit with stunning detail.

Such a man would have to be notoriously vocal and have deliberate speech. He would have to use his intellect to motivate, teach, and move the masses into action with his methodical—and logical—evolutionary ways. In fact, he might sound something like this:

> **Since the inferior always outnumber the superior, the former would always increase more rapidly if they possessed the same capacities for survival and for the procreation of their kind; and the final consequence would be that the best in quality would be forced to recede into the background. Therefore a corrective measure in favor of the better quality must intervene. Nature supplies this by establishing rigorous conditions of life to which the weaker will have to submit and will thereby be numerically restricted.**[1]

> Therefore, **the struggle for the daily livelihood leaves behind in the ruck everything that is weak or diseased or wavering; while the fight of the male to possess the female gives to the strongest the right, or at least the**

possibility, to propagate its kind. And this struggle is a means of furthering the health and powers of resistance in the species. Thus it is one of the causes underlying the process of development towards a higher quality of being.[2] That is why, *generally, the most vigorous males, those which are best fitted for their places in nature, will leave most progeny.*[3]

But in many cases, victory will depend not on general vigour, but on having special weapons, confined to the male sex.[4] Therefore, *can we doubt (remembering that many more individuals are born than can possibly survive) that individuals having any advantage, however slight, over others, would have the best chance of surviving and of procreating their kind?*[5] But such a preservation goes hand-in-hand with the inexorable law that it is the strongest and the best who must triumph and that they have the right to endure. He who would live must fight. He who does not wish to fight in this world, where permanent struggle is the law of life, has not the right to exist.[6] Evolution demands, *let the strongest live and the weakest die.*[7]

The stronger must dominate and not mate with the weaker, which would signify the sacrifice of its own higher nature. Only the born weakling can look upon this principle as cruel, and if he does so, it is merely because he is of a feebler nature and narrower mind; for if such a law did not direct the process of evolution then the higher development of organic life would not be conceivable at all.[8]

When we reflect on this struggle, we may console ourselves with the full belief, that the war of nature is not incessant, that no fear is felt, that death is generally prompt, and that the vigorous, the healthy, and the happy survive and multiply.[9]

Great Charles' ghost! My dear friends of science, readers of philosophy, men of deep thought, and haters of faith, hear my words now.

I must say that I wish I had written the tremendous speech above that captures the essence of evolution in a way that only pure genius can. Does it not envelope the complete spirit of evolution, natural selection, and survival of the fittest? It speaks truth to the dominant race: we have the right to struggle, fight, and subdue. It provides impassioned motivation to enact the Sub-Laws into our everyday lives. In doing so, we can become the very venom of evolution and paralyze the weak as they fall to the strong.

Actually, this speech is a compilation of words from the two most important evolutionists this world has ever seen—the father of evolution and his prodigal son. One was the creator of evolution, the other was the one strong enough to inflict it upon every living being he could touch.

Who were these great men? One was Charles Darwin and the other was his prized pupil, Adolf Hitler. Hitler's words are in **bold** and Darwin's are in *italics*. The words I wrote to piece their thoughts together are in regular type. I marvel at how closely Adolf Hitler adopted not only the concepts of Charles Darwin, but embodied his spirit and style, as if he wrote those words with a pen he robbed from Darwin's grave.

These men were two peas out of the same pod, a hand in a glove. They said the same thing, in such a similar fashion, that they sounded like the same person. When I researched these two individuals, I was beside myself; I have seldom come across one individual who so perfectly mirrored the will and testament of another. Yet, here we see Adolf Hitler doing just that of his intellectual and philosophical mentor, Charles Darwin.

If you are a true evolutionist, you should not be surprised at this revelation, for the writing has long been on the wall—you just couldn't decipher it. Allow me to translate the hieroglyphics for you. I will take you back in time and divulge what the history books have omitted.

For whatever reason, the ties between Adolf Hitler and Darwin have been kept under lock and key until recently when a few bold persons, including myself, cracked the lock on the safe. I admit that I, Dr. Iman Oxidant, am not the first person to jump into this pool, but rest assured that I will make the biggest splash. Here comes the cannonball.

Have you ever heard of eugenics? If not, that is a pity. Allow me to enlighten you:

> **eugenics:** *the science of improving a human population by controlled breeding to increase the occurrence of desirable heritable characteristics.*
> — NEW OXFORD AMERICAN DICTIONARY

A man by the name of Francis Galton largely developed the idea of eugenics. Galton was the cousin of Charles Darwin; they shared the same grandparent, Erasmus Darwin. Galton was born on February 16, 1822, and died January 17, 1911, just two years after he was knighted. He was considered a child prodigy, and like other well-known great men of the time, he was born into a wealthy family. In his application of scientific principles, we can see that Galton and Darwin were closely related in thought, as well as blood.

Galton coined the term "eugenics" which comes from a Greek stem meaning "good in birth."

> *Sir Francis Galton, a cousin of Charles Darwin, theorized that if talented people married only other talented people, the result would be measurably better offspring.*[10]
> — EDWIN BLACK

Galton developed the idea of selective breeding in humans from drawing upon the evolutionary doctrine of his cousin, Charles Darwin. But the family connection to eugenics does not stop with the cousin of Darwin. Major Leonard Darwin, Charles Darwin's very own son, succeeded Galton as the Chairman of the British Eugenics Society and served from 1911-1928. Major Leonard Darwin was a keynote speaker for the first International Eugenics Conference in London in 1912. He also spoke at the

Second International Eugenics Conference held in New York City in 1921.

Darwin's speech in New York was covered in the *New York Times* on September 25, 1921, in an article entitled, "Want More Babies in Best Families." The subtitle read, "Major Darwin Sees It Patriotic Duty of Better Classes to Increase Their Offspring— LIMITATION ALSO NEEDED—Danger of Best Types Disappearing and the Inferior Multiplying, He Tells Eugenists."[11] (The young impressionable Adolf Hitler must have felt that perfect scratch for his itching ears.)

Those of you who are new to eugenics will see that it was very much like our Sub-Laws of Evolution. It was the honest application of evolutionary principles to humanity.

LIKE A TREE
EUGENICS DRAWS ITS MATERIALS FROM MANY SOURCES AND ORGANIZES THEM INTO AN HARMONIOUS ENTITY.

The logo of the Eugenics movement is quite interesting; it even seems to sport Darwin's Tree of Life.[12]

As you can see, "Eugenics is the self-direction of human evolution." The movement took Charles Darwin's evolution and applied it to humans.

In the early nineteen hundreds the eugenics movement picked up steam and crossed the pond. California became the center of the eugenics movement in the U.S. and Americans followed Galton and Darwin and implemented evolutionary ideas.

The work of these pioneers was quite spectacular; American citizens of today may be surprised to find out how much progress was actually made in the beginning of the twentieth century toward the adoption of evolution into the human lifestyle. Even more surprising to some are the powerful names behind the movement; some of these families still have tremendous influence in the U.S. government today.

> *Eugenics would have been so much bizarre parlor talk had it not been for extensive financing by corporate philanthropies, specifically the Carnegie Institution, the Rockefeller Foundation and the Harriman railroad fortune. They were all in league with some of America's most respected scientists from such prestigious universities as Stanford, Yale, Harvard and Princeton.*[13]
>
> — EDWIN BLACK

Early on in eugenics, esteemed educational institutions—that still unapologetically adhere to evolution—were the brains behind evolutionary eugenics. They were on the ground floor assisting those who sought to make evolution a lifestyle for humans. As always, we can count on educational institutions to stay true to their colors. Without them, I don't know where evolution would be today.

Early on, educational institutions and scientists were not alone in their quest for evolutionary perfection. Schools were the brains, and the powerful supplied the money, but someone else had to be the brawn, as we can see by the continued help Edwin Black cited in his article, "Eugenics and the Nazis—California Connection":

> *Even the U.S. Supreme Court endorsed aspects of eugenics. In its infamous 1927 decision, Supreme Court Justice Oliver Wendell Holmes wrote, "It is better for all the world, if instead of waiting to execute degenerate offspring for crime, or to let them starve for their imbecility, society can prevent those who are manifestly unfit from continuing their kind...Three generations of imbeciles are*

125

enough." This decision opened the floodgates for thousands to be coercively sterilized or otherwise persecuted as subhuman. Years later, the Nazis at the Nuremberg trials quoted Holmes' words in their own defense.[14]

— EDWIN BLACK

Elements of the philosophy were enshrined as national policy by forced sterilization and segregation laws, as well as marriage restrictions, enacted in 27 states. In 1909, California became the third state to adopt such laws. Ultimately, eugenics practitioners coercively sterilized some 60,000 Americans, barred the marriage of thousands, forcibly segregated thousands in "colonies," and persecuted untold numbers in ways we are just learning.[15]

— EDWIN BLACK

I just noticed that peculiar little word—segregation. I find it ironic that religion has carried the blame for segregation when it was, in fact, science that won over the people in power who had the authority to enforce segregation. Indeed, we have done well in blaming religion for segregation, and consequently bigotry, when in fact, bigotry was the mistress who slept in the bed of evolution:

Creation myths were in a sense the beginning of science itself. Fabricating them was the best the early scribes could do to explain the universe and human existence. Yet the high risk is the ease with which alliances between religions and tribalism are made. Then comes bigotry and the dehumanization of infidels.[16]

— E.O. WILSON

E.O. Wilson was elaborating on the foundations of religion. He then wrote that religion leads to bigotry. I tell you, that is like the teapot calling the kettle black; it makes me chuckle every time we pull that one off.

I hope some of you have realized by now that science does not only take place in labs. Science has a history of, and a ripe future in, forcing the hand of government and culture and using them to enforce our tenets. Evolutionary science, through eugenics, played a critical role in government policy in the earlier part of the last century, and today, it is critical in government policy in our school curriculum.

We depend on government, for example, to fight our battle against the notion of intelligent design. Having our political forces determine what is science and what is not has been one of our most useful tools in recent times. Philosophy professor Elliot Sober uses such a tactic in his article, "What is Wrong With Intelligent Design":

> One reason is that versions of creationism that mention a supernatural being have a Constitutional problem—U.S. courts have deemed them religious, and so they are not permitted in public school science curricula.[17]

In his article, Sober makes a good point. He reasoned that since the U.S. Constitution does not allow the teaching of creationism in schools, then it must not be science. Therefore, intelligent design is not valid.

I am not sure when we went to the Constitution to teach us about the origins of organic life, but it stuck to the wall, so we will keep throwing it. We will use any method we can to promote our evolutionary agenda, even government. I understand that government is not science, but if we can allow laws to defeat the enemies of our position, then our work will be easier, we won't have to scientifically defend our position, and we will be looked upon with less scrutiny.

In fact, the following is one of my favorite examples of this concept. A school board in Georgia placed a sticker on the cover of a science textbook stating that evolution is a theory, not a fact. Proponents of evolution sued the school system, even though at the time evolution *was* called the "theory of evolution."

I would like to add a keen observation, if I may. In the second sentence the sticker[18] should have stated, "regarding the origin of living things and the preservation of the favored races." I am

> This textbook contains material on evolution. Evolution is a theory, not a fact, regarding the origin of living things. This material should be approached with an open mind, studied carefully, and critically considered.
>
> *Approved by*
> *Cobb County Board of Education*
> *Thursday, March 28, 2002*

sure if the Cobb County Board of Education had printed that, the evolutionists would not have sued. But they did not, and some angry parents who believed in evolution, took the battle to court in a noble attempt to obtain justice.

If we could conduct our own analysis, we might be able to get to the bottom of why this sticker would have done much harm had it been allowed to remain in the science textbooks. While the sticker does nothing more than ask science students to study carefully, with an open mind, and critically consider evolution, I believe that is precisely where the school board found the trouble that triggered the lawsuit.

As scientists and evolutionists will tell you, studying evolution carefully, with an open mind, and using critical thinking, is a gigantic waste of time. And if you do that, we will take legal action. If we wanted students to study carefully like the sticker insinuated, I doubt the lawsuit would have ever been initiated.

On the contrary, the outcome of this lawsuit sums up all evolutionary thought superbly. Those who are less-evolved cannot think, cannot be trusted to have open minds, and do not have the credentials to be critical of what the elite evolutionists say, write, or teach. We will spoon feed you the truth—and you will like it. And, if you try to think, study, and have an open mind, there are people who will try to sue your derrieres right off.

Thankfully, the evolutionists came out on top when the sticker nonsense was finally settled. The following is the outcome of the case as stated in a Cobb County School Press release:

> *Under the agreement, the District will not attempt to place the same, or similar, stickers in textbooks again. In return, plaintiffs have agreed to end all legal action against the school district. In a separate agreement, the District has agreed to pay $166,659, which represents a portion of the plaintiff's legal fees.*[19]

Not only will some oppose you if you advocate open-mindedness, you may even pay the legal fees of the evolutionists; after all, you caused the lawsuit by ruffling the internal balances of our evolutionary minds. I dare say no school district in our nation will challenge us again. The writing is on the wall; we will fight and you will lose.

Now, mind you, I don't want to lead anyone astray. Science has historically encouraged critical thinking, studying, and open-mindedness. In fact, some might even go so far as to say such attributes are the backbone of good science. However, they are just not the backbone of evolution, as evidenced by this court case. The facts show that a lawsuit was filed when a rogue school challenged students, via a sticker, to study with an open mind and to use critical thinking when studying evolution.

With that boost of adrenaline, let's press on.

As evolutionists, we admit there is evidence to support harboring racist feelings. As we have already ascertained, it is an evolutionary fact that some races are superior to others. In modern cultures, however, expressing feelings of anti-Semitism, supporting segregation, and using racial slurs can sometimes be detrimental to our movement.

In such times, we have been able to blame churches and religion for the source of those feelings. As mentioned earlier, E.O. Wilson said religion is the source of bigotry. A brilliant move I must say. Way to divert the bullet!

Early in the nineteen hundreds, America was following Darwin down his road, but unfortunately for our country, another nation was moving forward with faster progress. It was this other nation's commitment to true evolutionary practice that changed the shape of evolution forever.

In 1934, as Germany's sterilizations were accelerating beyond 5,000 per month, the California eugenics leader C. M. Goethe, upon returning from Germany, ebulliently bragged to a colleague, "You will be interested to know that your work has played a powerful part in shaping the opinions of the group of intellectuals who are behind Hitler in this epoch-making program. Everywhere I sensed that their opinions have been tremendously stimulated by American thought . . . I want you, my dear friend, to carry this thought with you for the rest of your life, that you have really jolted into action a great government of 60 million people." [20]

— EDWIN BLACK

That same year, 10 years after Virginia passed its sterilization act, Joseph DeJarnette, superintendent of Virginia's Western State Hospital, observed in the Richmond Times-Dispatch, "The Germans are beating us at our own game." [21]

— EDWIN BLACK

I mentioned earlier that evolutionists stick together through thick and thin; the bond of evolution is even thicker than nationality, for we are bound by racial blood. Edwin Black wrote that "The Rockefeller Foundation helped found the German eugenics program and even funded the program that Josef Mengele worked in before he went to Auschwitz."[22]

Hitler studied evolution, and the most powerful families, scientists, and institutions in Europe and the United States helped him do it. Hitler used eugenics, purely Darwinian logic, to build his empire of Aryan (white) supremacy. Hitler's love affair with Darwin's idea of "favored races" was a real and long one; Hitler knew you couldn't fight science.

Hitler even wrote a fan letter to American eugenics leader Madison Grant, calling his race-based eugenics book, "The Passing of the Great Race," his "bible." [23]

— EDWIN BLACK

Some might say Hitler perverted evolution, but that is not the case. He perfected it! Even American evolutionists believed Hitler was leading the way and was right on target:

> *While we were pussy-footing around, the Germans were calling a spade a spade.*[24]
>
> — LEON WHITNEY,
> EXECUTIVE SECRETARY OF THE AMERICAN EUGENICS SOCIETY

The fact is, Hitler was simply an overachieving evolutionist, and we should honor him as such. While Charles Darwin did not say specifically to invade France and kill Jews, he penned a worldview that allowed, affirmed, and even supported those actions in the name of natural selection and survival of the fittest.

Arguing whether evolution encouraged Hitler to do what he did is a moot point because evolution allows, and approves of what he did. When it comes to Nazi Germany's relationship with the Jews, it was nothing more than a stronger group of people pressing on a weaker group and forcing its destruction. That, my friends, could not be any more Darwinian.

> *One large group will slowly conquer another large group, reduce its numbers, and thus lessen its chance of further variation and improvement. Within the same large group, the later and more highly perfected sub-groups, from branching out and seizing on many new places in the polity of Nature, will constantly tend to supplant and destroy the earlier and less improved sub-groups.*[25]
>
> — CHARLES DARWIN

> *If any one species does not become modified and improved in a corresponding degree with its competitors, it will soon be exterminated.*[26]
>
> — CHARLES DARWIN

Adolph Hitler had the support of Darwin and evolution, and if you are a true evolutionist, he should have your support as well.

Hitler believed the Jews were a less-evolved, lower form of life. This belief allowed him the latitude to press upon and exterminate the Jews if his race was indeed the stronger race. As history shows, many Jews fell to his Survival of the Fittest aggression.

Hitler's gavel swung in the hallowed halls of evolution and it could be heard throughout the world; it was in the gas chambers and in the ovens Hitler used to exterminate his lesser-modified opponents. One could almost hear the Nazis decree evolutionary logic perfectly as the oven doors slammed shut: "Let the strong live and the weakest die."

All Charles Darwin would have said about the Holocaust is that it was the quintessential struggle for life, and through that struggle, only the strong survived. Only the strong should have the opportunity—and the right—to procreate their kind. Darwin would have said to the Jew:

> *When we reflect on this struggle, we may console ourselves with the full belief...that no fear is felt, that death is generally prompt, and that the vigorous, the healthy, and the happy survive and multiply.*[27]
>
> — CHARLES DARWIN

Touché.

Hitler had the support of scientists as well as some of the most powerful Americans. It has even been asserted that Winston Churchill was a believer in eugenics, and that he admired Hitler's rapid rise to power. Now, Churchill's tone might have changed when his country was being bombed and nearly 400,000 of his citizens were killed, but who's to say.

One of the methods Hitler used to create his evolutionary society was to give power to the doctors and scientists. That is a move we see our progressive societies doing today, as we discussed in the "Reduction" seminar. Needless to say, evolutionists understand that doctors and scientists who consistently apply evolutionary doctrines are above reproach, for they carry the best intentions of mankind with them at all times—even if society isn't quite ready for what has to be done.

> *Eugenic breeders believed American society was not ready to implement an organized lethal solution. But many mental institutions and doctors practiced improvised medical lethality and passive euthanasia on their own. One institution in Lincoln, Illinois, fed its incoming patients milk from tubercular cows believing a eugenically strong individual would be immune. Thirty to forty percent annual death rates resulted at Lincoln. Some doctors practiced passive eugenicide one newborn infant at a time. Other doctors at mental institutions engaged in lethal neglect.*[28]
>
> — EDWIN BLACK

History shows us that Hitler's doctors were essential to his goals as well.

> *Nazi doctors would become the unseen generals in Hitler's war against the Jews and other Europeans deemed inferior. Doctors would create the science, devise the eugenic formulas, and hand-select the victims for sterilization, euthanasia and mass extermination.*[29]
>
> — EDWIN BLACK

They were on the ball, I tell you. Today, our doctors are holding firm to their proud roots; they are utilizing the syringe of abortion and the fatal poison of euthanasia. Doctors and scientists are the muscle behind the movement. Hail to the abandonment of the Hippocratic Oath.

In addition to exterminating the Jews, Hitler embraced more from evolutionary thought. Any individual who was deemed inferior was killed, regardless of race. In this model, only the strongest individuals of the strongest race were allowed to live and breed. That is good evolutionary strategy and is very much in line with our Sub-Laws of Evolution.

The official name of Hitler's evolutionary euthanasia program was Action T4, but it was also called Operation T4 as noted here by D. Ryan and J. Schuchman in *Deaf People in Hitler's Europe*:

In all, historians estimate that 200,000-250,000 institutionalized mentally and physically disabled persons were murdered under Operation T4 and its corollaries between 1939 and 1945.[30]

This history is well documented and we know Germany lost the war, so these programs were relatively short-lived. In the end, the Nazi officials who ran these programs and extermination camps were put on trial during the famous Nuremberg Trials. Many were convicted for their exploits under the Nazi regime and for mass killings of Jews.

One point of extreme interest for our cause is that during the Nuremberg Trials of 1945, Nazis were also convicted and executed for killing those with birth defects (like blindness, schizophrenia, and other offensive physical deformities as outlined in Action T4). Just sixty-one years later, however, in the year 2006, the good Church of England and Bishop Tom Butler of Southwark publicly announced that it is perfectly fine to kill people with birth defects.

How remarkable the human species really is! One generation called the act of euthanasia an atrocity punishable by death, and the very next calls the same act compassionate and humane. This is also evident in the progressive societies in Europe that promote euthanasia of many forms. The teeth of evolution have definitely sunk into society. We are winning my good friends; we can see the Fruit of Evolution sprouting up everywhere! And it has only just begun.

One cannot underestimate the power of our movement. The Church is our enemy, yet even they do our bidding by trying to blend the truths of science with their feeble scriptures. We now have church leaders calling for the death of children with birth defects. That proves our Tree of Life is growing strong and setting roots, for that very same act, sixty-one years earlier, was considered the crime of the century. Churches condemned the Nazi onslaught and the world watched in horror at their heinous atrocities and applauded the executions of the perpetrators.

Did God change his mind, or do people finally understand that there is no standard set by an imaginary God? Standards are only set by man—as evidenced by the societal shift we have seen. As Bishop Butler has shown us, even the Church's standards are about as solid as lukewarm water.

We can credit this shift to the power of relentless science, my friends. For you see, the Nazis disappeared, and the term "eugenics" was ousted from the rhetoric of scientists, teachers, and doctors. But only the term was eliminated; eugenics is nothing more than evolution applied to humans. Even though we have stopped talking specifically about eugenics, it remains mandatory that evolution, the architecture of eugenics, is taught to every student in the public school system. And, indeed, eugenics is practiced today as parents are counseled to abort those fetuses with genetic deficiencies. Eugenics is alive and well, though somewhat disguised in our society.

This was not the case at the end of the World War II when the United States still had a religious culture. Our culture today, however, is a product of our evolutionary indoctrination and our relentless attack on Christianity through the schools and the government. It is this surge—one that neither tires nor waivers— that must be given the credit for this wonderful societal shift and the mass complacency that allows doctors to perform abortions and euthanasia.

As we continue to break down faith in God and replace it with evolutionary thought, societies worldwide will be desensitized to abortion and the euthanizing of infants. At that point, we can make the logical argument, just like Hitler did, that if we destroy an infant for a physical defect, why can't we use euthanasia techniques on someone just five years older—then ten, then thirty? Then the floodgates will be open for anyone to be euthanized for any physical or mental defect. Since February 2007, Switzerland has allowed the euthanizing of people with mental illness. So we will be supporting them and watching their progress very closely.

Right now we are in the middle of a huge stride forward in living life within the rules of evolution, and as a result, our world will be a better place. Our population hawks may even get their wish to see 5.85 billion people die. If you love humanity, you should advocate that death toll as well. It is the only way for us to survive.

On that note of eternal hope, can you imagine what could happen in the next sixty years? If we replicate the societal shift we have seen in the past sixty years, we should be able to use euthanasia procedures on anyone, for any defect, even the defect of having faith in God.

At birth we exhibit many traits, which over time, may prove to be physical and mental defects. As doctors become more sophisticated in diagnosis they will be able to identify nearly every physical and mental disorder in humans very early in life. In an attempt to improve the racial stock, evolution will allow us to sift the herd and eliminate those defects swiftly and thoroughly.

One of those defects is faith, and we will be able to move against religion in a very deliberate way. As we have learned, God is a myth, and faith in him is delusional. Faith in God, therefore, is a mental disorder just like schizophrenia. Because scientists, doctors, and judges can do no wrong, we can count on them to provide options for aborting, sterilizing, and exterminating the degenerate of our species. That includes delusional religious zealots.

All it takes is for another Darwinian leader to rise up, one with the strength of the favored race in his blood, a scientific education, a concern for the welfare of the planet, charisma that shines from north to south, and a birthright of power, so that his agenda can be implemented. We need another Adolf—only better. We need someone who can inspire, captivate, and motivate the world, even those he plans to exterminate. We need a king.

My dear students, these seminars that have been sponsored by The Institute of Progressive Lineage, are now coming to a close. We have scripted the Sub-Laws of Evolution, and partaken of its very Fruit. I realize that for some—all women, all inferior races, those of faith, the handicapped and defective, those forcibly impregnated, and those killed for their possessions—the Fruit may have a slightly bitter taste. You can rest assured that it is nature itself that has dictated the predicament in which you find yourself.

Whether you are African, South American, Eskimo, Aborigine, American Indian, or some other less-evolved race, you should be grateful you have lived this long. Be thankful for the short-lived compassion of the "favored race," for there is no doubt we have allowed you to survive longer than you should have. So in this, rejoice. You have had a nice run, but the strong will ultimately cause much destruction to the weak. This is the way of Darwin's evolution.

I know it might be a sore spot knowing your children are being taught evolution in schools, when the very doctrines they are learning point out their inferiority, but you can't fight science.

A message of caution for those who have daughters who are of the favored race: with our new training and understanding in Darwinian evolution, we are instructing our young men of the favored race to take and breed favored-race girls so their superior DNA can be replicated and fill the Earth as opposed to the weaker strands of humans. If you do not wish your daughters to be impregnated by these dominant males, I would certainly advise some form of birth control.

To fully adopt the evolutionary lifestyle, we will need help facilitating the reproduction of superior genetics and eliminating the inferior populations. We must restrict wretched and "miserable creatures" from "breeding their brains out," as we borrow language from Charles Darwin and Dr. Eric Pianka, respectively.

I can finally confess that my wisdom is exhausted. Yet, I would not be so presumptuous as to end our legendary seminars in my own words, even though they resound around the globe. I leave you with the words of our greatest fathers of evolutionary thought and practice:

> *He who would live must fight. He who does not wish to fight in this world, where permanent struggle is the law of life, has not the right to exist. Such a saying may sound hard; but, after all, that is how the matter really stands.*[31]
> — ADOLF HITLER

Only Charles Darwin explained evolution better: *Let the strongest live and the weakest die.*[32]

On that note, I bid you farewell. Bottoms up from the bunker below the cellar in Boston, Massachusetts.

Most tender warm regards,

Dr. Iman Oxidant
Chairman of The Institute of Progressive Lineage

ENDNOTES

1   Adolf Hitler, *Mein Kampf*; First Volume, Hurst and Blackett Ltd, 1939, pg. 161~162

2   Ibid., pg. 161

3   Darwin, Charles, *From So Simple a Beginning*, New York: W.W. Norton & Company, Inc., 2006, 506

4   Ibid., pg. 506

5   Ibid., pg. 502

6   Adolf Hitler, *Mein Kampf*; First Volume, Hurst and Blackett Ltd, 1939, pg. 163

7   Darwin, Charles, *From So Simple a Beginning*, New York: W.W. Norton & Company, Inc., 2006, pg. 605

8   Adolf Hitler, *Mein Kampf*; First Volume, Hurst and Blackett Ltd, 1939, pg. 161

9   Darwin, Charles, *From So Simple a Beginning*, New York: W.W. Norton & Company, Inc., 2006, pg. 605

10 Black, Edwin, "Eugenics and the Nazis—The California Connection," *San Francisco Chronicle*, November 9, 2003.

11 "Want More Babies in Best Families, *The New York Times*, September 25, 1921.

12 Exhibit photograph scanned from: Harry H. Laughlin, *The Second*

International Exhibition of Eugenics held September 22 to October 22, 1921, in connection with the Second International Congress of Eugenics in the American Museum of Natural History, New York (Baltimore: William & Wilkins Co., 1923).

13 Black, Edwin, "Eugenics and the Nazis—The California Connection," *San Francisco Chronicle*, November 9, 2003.

14 Ibid.

15 Ibid.

16 Darwin, Charles, *From So Simple a Beginning*, New York: W.W. Norton & Company, Inc., 2006, pg. 1483.

17 Elliot Sober, "What Is Wrong With Intelligent Design?" *The Quarterly Review of Biology*, March 2007.

18 Sticker can be viewed at, among others: http://www.msnbc.msn.com/ID/6822028/.

19 Cobb County School District Press Release, "Agreement Ends Textbook Sticker Case," December 19, 2006.

20 Black, Edwin, "Eugenics and the Nazis—The California Connection," *San Francisco Chronicle*, November 9, 2003.

21 Ibid.

22 Ibid.

23 Ibid.

24 Ibid.

25 Darwin, Charles, *From So Simple a Beginning*, New York: W.W. Norton & Company, Inc., 2006, pgs. 530-531.

26 Ibid., pg. 515.

27 Ibid., pg. 500.

28 Black, Edwin, "Eugenics and the Nazis—The California Connection," *San Francisco Chronicle*, November 9, 2003.

29 Ibid.

30 Donna F. Ryan, John S. Schuchman, *Deaf People in Hitler's Europe*, pg. 62.

31 Adolf Hitler, *Mein Kampf*, Vol. 1, Chapter 11, pars. 25-26.

32 Darwin, Charles, *From So Simple a Beginning*, New York: W.W. Norton & Company, Inc., 2006, pg. 605.

CHAPTER FOURTEEN

THE SAMARITAN TRAIT

Evolution is a complex web of deceit. It takes a deliberate illustration, without interruption, to untangle that web. That is why I created the fictional character Dr. Iman Oxidant.

I believed if I simply soaked the reader in pure evolutionary thought without pause or reprieve, that evolution would drown itself. For as it has been shown, a life lived by the sole standards of natural selection is not a sustainable life and it's impossible to fashion a society around it.

It is a system of combat, collision, and claw. In its most honest form, evolution governs life only through thievery and death. It represents the most frightening environment imaginable where hope could only be a cruel and empty emotion.

It is undeniable that Hitler carried evolution to its appropriate end: death and mayhem guided by no trace of conscience. Nazi Germany was secularism run unchecked without any moral hesitation to impede it. Hitler believed the strong should live and the weak should die. He believed in natural selection in its purest form and he was capable in enforcing it. The result was the entire world engulfed in war.

Do not be convinced that Hitler was a rogue and perverted the ideologies of Darwin; he was the epitome of evolutionary implementation. Hitler was backed by the writings of Charles Darwin, and did nothing wrong if measured against the standards of natural selection. Based on survival of the fittest, the strong have every right to exterminate, extinguish, sterilize, rape, and murder the weak.

Over forty-eight million souls were killed during World War II because of Hitler's aggression and the spine given to him through Darwin's infamous book, *On the Origin of Species Through the Means of Natural Selection or the Preservation of Favored Races in the Struggle for Life.*

If evolution through natural selection were true, and humanity was formed and born out of its forces, then we would all live in a perpetual state of world war without reprieve. Only the strongest would be allowed to survive. And of course, those who are strong would always be in flux because age impacts physical strength. It would be a reign of violence where when one bleeds, one dies.

But we don't live in a perpetual war where each day is a struggle for life. We are a people of compassion and give aid to those in need. We are not driven by cold animalistic instinct. We are humane and demonstrate love.

Even atheists live by a code of conduct that is completely foreign to their own belief system. The evolutionists like to say they are a product of evolution, but when it comes down to it, they really don't live by the iron fist of survival of the fittest. They benefit from morality-based laws that make survival of the fittest mentalities illegal and punishable.

The mechanism of natural selection, as the sole source of forming and evolving humans, can only be seen in textbooks. It surely cannot be seen at the Salvation Army, food drives, foreign aid to fight the AIDS epidemic in Africa, or any other "humanitarian" effort that serves, aids, and protects those in need.

It is the atheist Richard Dawkins who said that the desire to adopt children must be a "genetic mistake."[1] He said this because he realized there is no logical room within natural selection to account for this kind of compassion for the offspring of another human. His evolutionary argument is that humans should be focused on perpetuating their own genetic line, not someone else's. Evan Eisenberg writes in the "The Adoption Paradox":

> *From a Darwinian standpoint, going childless by choice is hard enough to explain, but adoption, as the arch-*

Darwinist Richard Dawkins notes, is a double whammy.
Not only do you reduce, or at least fail to increase, your own
reproductive success, but you improve someone else's. Since
the birth parent is your rival in the great genetic steeplechase,
a gene that encourages adoption should be knocked out of
the running in fairly short order.[2]

Under evolution, this is absolutely correct. Adoption makes
no sense. The desire to adopt children would have never surfaced
through natural selection—yet another inconsistency between the
reality of how humans live, and how humans ought to live, if we
were a product of evolution.

For evolutionists, this incompatibility pointed out by
Dawkins is a much larger problem than you might think.
Evolutionist Maxine Singer, president of the Carnegie Institution
of Washington, noted in a column in the *Washington Post* that the
word "theory" has a special meaning:

A theory in science is not a hunch or "just a theory"
as some say. It is an explanation built on multitudinous
confirmed facts and the absence of incompatible facts.[3]

According to Singer, a scientific theory has no incompatible
fact held against it. Dawkins however, pointed out an incompatible
fact. What he pointed out was much deeper than just someone
wanting to adopt a child, which is a peculiar occurrence that
should not be happening.

The deeper issue Dawkins unwittingly conceded is that human
nature and human habit do not resemble what a product of natural
selection should be. Dawkins clearly unveiled—through his views
on adoption—that evolution through natural selection fails in
its explanatory powers when one invokes a simple observational
study on the habits and lifestyles of the human species. Simply
put, Dawkins understands that natural selection cannot explain
why we are the way we are. Yet, that is the stated intention of the
theory.

This is just one of the areas where the theory of evolution fails. Darwin observed the tooth and claw nature of animal life and then applied his conclusions about animals to humans. He understood that for evolution to work, the method of natural selection had to be all encompassing on all life forms. Therefore, humans were carelessly included in the theory even though animalistic instinct does not directly translate to the lives of humans.

It would have been impossible to formulate the theory of evolution through natural selection if the study group under observation were solely humans. That is why evolutionist Eric Pianka was correct when he stated that humans break all the rules (he was speaking from the mind of an evolutionist). Our actions, our thoughts, our mentalities simply do not fit into this system they have created to explain us.

In contrast, the world actually resembles the Bible's account, in which humans are independently made, given dominion, and bestowed an elevated role where our decisions impact the entire globe—for better or for worse. We have capabilities for great evil, yet through grace, we have capabilities for great good. It is the latter that evolution cannot explain.

Atheist P.Z. Myers has this to say about this inconsistency of human habit. He does so through a book review of Dawkins' work, *The God Delusion,* in *SEED Magazine* in which Dawkins tries to deal with morality:

> *Within this framework Dawkins* considers the problem with morality. *Moral behavior is clearly not an inevitable result of religion—which seems to provoke more barbarism than it prevents—*but it's also not immediately obvious how a Darwinian regime would foster kindness and charity.[4] *(Emphasis added)*

Myers clearly recognizes the shortcomings of evolution as an explanation of real humanity. However, the most interesting point to me is the beginning part of the second sentence before he concedes that evolution draws a blank on kindness and charity.

Myers, knowing evolution is about to take one on the chin, throws a wild haymaker toward religion. It is a panicked attempt to remind the reader that religion still has nothing to offer, even though his brand of science cannot figure this one out.

He continues to discuss Dawkins' thoughts on the subject:

> *While considering the ideas of kin selection and reciprocal altruism as reasonable mechanisms for introducing moral behavior, Dawkins again proposes that a wider morality, applied to more than just relatives and trusted friends, is a byproduct, an accidental gift of nonspecific feelings of generosity or sympathy.*[5]

Again we see that anything good in our character is merely an accidental gift that does not make any sense for us to have. Instead of my words of rebuttal, however, see how Myers disciplines the reasoning of Dawkins:

> *But I think we need a little something more to maintain altruism than to call it a lucky mistake—self sacrifice for the benefit of unrelated individuals ought to be selected against.*[6]

What Myers points out is the sheer fact of the matter. The Samaritan Trait, which is the human tendency to selflessly help those in need, is widespread. The presence of this selfless compassion should not have been developed in humanity under the system of natural selection. As Myers states, "self sacrifice... ought to be selected against."[6]

However, displays of human compassion and self sacrifice are an observable, repeatable condition found at every age, in every culture, every country, and every race, and it is a screaming contradiction to natural selection—the alleged driving force of evolution.

Myers, after dismissing Dawkins' ideas, does offer his own philosophical reason for the evolution of morality, which he states, could be "empathy." He contends that such empathy would produce the "golden rule."

First of all, I find it most revealing that atheistic evolutionists must look to Jesus to find a standard by which to live. It is a clear sign that within their belief system there is no governing morality that gives them the security that they need.

The Golden Rule is in the Sermon on the Mount cited in Matthew 7:12:

> So whatever you wish that others would do to you, do also to them, for this is the Law and the Prophets.

Myers is not alone in citing the Golden Rule; Dawkins has also invoked it as the way we should all want to live.

Secondly, if I were an evolutionist needing a solid explanation on why humans act the way we do, I can't even imagine trying to intellectually reconcile with myself that perhaps empathy is the reason why the entire human species has morality.

Furthermore, it almost goes without saying, that there is no evidence to back up either Myers' "empathy," or Dawkins' "accidental gift" idea that could actually prove their explanation was the cause of our species-wide trait of morality.

It is just a choking amount of meaningless philosophy resulting from trying to merge and swallow two opposites: survival of the fittest and compassion. Evolutionists continue to write checks their evidence can't cash, resulting in a bankrupt ideology where every good character trait in mankind becomes the most peculiar, stupefying phenomenon for which there is no reasonable explanation.

Through the words of two prominent and vocal evolutionists, evolution through natural selection is incompatible and insufficient as an explanation of our humanity. Add Maxine Singer's definition of a theory to the mix, and we see the complete failure of evolution through natural selection. A true scientific theory has no opposing fact held against it, and evolution through natural selection does— the Samaritan Trait found in humanity.

There is a deeper truth that is inherent in all of this. It is not a big revelation to many of us that humanity is not a product of evolution through natural selection. God's hand and plan have been evident since the beginning of this world. In the same way, there will always be opposition to accepting a Creator. That is the result of the original sin.

Yet, in understanding the pitfalls of evolution, we stand on one of the most profound doctrines of truth; one that impacts us on such a deep level that it can be overwhelming to grasp its fullness. It is this: evolution is completely disbarred and dismantled when we are Christ-like.

Never has the saying, "actions speak louder than words" meant more than it does right now. Natural selection/survival of the fittest is shattered as a plausible explanation to human life when people exhibit the fruit of the Spirit!

> *But the fruit the Spirit is love, joy, peace, patience, kindness, goodness, faithfulness, gentleness, self control.*
> — GALATIANS 5:22-23

As Christians show the evidence of a heart changed by God, they embody the irreconcilable contradiction of evolution through natural selection. As possibly hundreds of millions have done before us over the last couple of millennia, when we live and produce the fruit of the Spirit, we prove that evolution through natural selection cannot be our origin.

When we exhibit Christ-like love, we display something that cannot be produced through the natural world as evidenced through Dawkins and Myers. The attributes of Christ that are visible in humanity, the greatest of them being love, become an unsolvable riddle for atheistic evolutionists.

Consequently, without repentance, atheists will always be confounded by the "problem with morality" and be hopelessly perplexed at the source of "kindness and charity." The powerful existence of the fruit of the Spirit that is perpetuated throughout mankind puts an indelible stamp of a greater power upon us.

Darwin's life became a monumental tale of tragedy. In the wake of his daughter's death, he refused to believe that if there was a God, that He could be a loving God. He then created an ideology where neither love nor God could exist. How much of this he realized at the time, I don't know.

Through biblical-view Christianity this truth is profound. We understand the fallen nature of mankind, which is vividly discussed in the Book of Romans. Therefore, we understand the only love we can truly show as humans, must come from God.

Darwin's theory reasons correctly that the natural world can offer no love, compassion, or refuge for the weak. It takes a supernatural force to intervene to create those attributes. However, since atheists do not believe in the supernatural, they cannot reconcile love or explain it, even though they can clearly and admittedly observe the footprints of love.

The existence of love in our lives makes evolution through natural selection obsolete in humanity. In a very tangible and literal way, love conquers all.

ENDNOTES

1   http://www.beliefnet.com/story/178/story_17889_2.html

2   Evan Eisenberg, "The Adoption Paradox." *Discover*, Jan. 2001.

3   http://whyfiles.org/095evolution/index.html, courtesy of University of Wisconsin Board of Regents

4   Myers, P. Z., "Bad Religion," *SEED Magazine*, November 2006, pg. 89

5   Ibid.

6   Ibid.

CHAPTER FIFTEEN

THE END GAME OF EVOLUTION

There is a deep disturbance in the underpinnings of evolution. When one fully grasps the reality of an ideology that has no answer or explanation, one that literally prohibits love, well, that should cause everyone to snap to attention.

It is this ideology of animalistic humanity that is being force-fed to our youth through the science curriculum in our public schools. Within this ideology, devoted parenting—particularly the act of adopting a child who is in need of unconditional love—is viewed as only a mistake. Abortion, a sick, heartless act, becomes the perfectly played hand in the game of survival of the fittest.

According to survival of the fittest, however, abortion is not the best answer for the mother because the mother should be interested in propagating her own seed and continuing her genetic line. Abortion is the answer for those who do not want her to give birth. Abortion is never about helping the mother; it is about protecting others from the annoyance of an unwanted child who would consume vital resources once outside of the womb.

Abortion comes from the mentality of "kill them while you can." Evolution, and everything that goes with it—accidental existence, valueless human life, uncertain futures, unstable environment, population overload and so on—allows its followers to feel *justified* in their abortion strategy.

That is why advertising campaigns that state, "abortion is murder" do not work. The people performing abortions understand what they are doing; they just have an ideology that allows them to do it. We cannot guilt them into remorse when there is no guilt.

As more and more layers of grime are scraped away from evolution, its core reveals a belief system that simply justifies the sinful cravings of people. For people to reject evolution, they would also have to reject the sin that engulfs their lives. Doing so is an act of repentance, which ultimately leads to salvation. Saving grace is truly the only answer to pulling those people who fully embrace evolution as their worldview away from its harmful ideology.

This is why the many scientific arguments against evolution are falling on deaf ears. Evolution has been portrayed as a science dealing with transitional skeletons, the age of the Earth, and the idea that humanity came from simpler beginnings. But we know that the scientific shell is only a sugarcoating on the pill. Before one accepts the science of evolution, one must first reject the biblical account of creation, and that in itself represents a decision to reject God. This denial of the biblical creation story is the prerequisite to evolution.

Theistic evolutionists (those who attempt to merge God and evolution) must still have a level of rejection as a prerequisite to evolution. They are choosing to reject the creation account of the Bible, which is the Word of God, for the story of evolution, which is the imagination of man. If one who is a born-again believer still holds onto evolution, their salvation is not in jeopardy, because God does not demand that we intellectually comprehend everything perfectly. But in their case, many of the storehouses of knowledge that are available are closed to them because of that rejection of Genesis. The correct foundation must be built upon to truly understand the world.

However, for most people, once that pill of evolution is swallowed it becomes much more than a science; it is the explanation of their entire existence. It is a religion that demands of them a lifestyle that is about unleashing their carnality and having no conscience to slow it down.

The religious characteristics of evolution are absolutely remarkable. Every year on February 12, there is a holiday called Darwin Day, a day of celebration and tribute to their hero who was born on that day in 1809. The people who observe this day call themselves "Darwinites," and some even go as far as to dress up like the "Father of Evolution" and wear fake beards. A statue of Darwin has been erected in his hometown of Shrewsbury, England.

Quite frankly, it looks and sounds like idol worship. As much as his followers like to say they are progressive, they are actually regressive. Their actions are like the people in the Old Testament who worshipped Baal, the golden calf, and other false gods. There is even a website asking for donations to support Darwin Day that contains proselytizing content.

The celebration of Darwin Day is not just a local parade to honor a town's homegrown celebrity. It is going global. Take a look at the following content—especially the title, which was taken from www.darwinday.org.

The Evolution of a Global Celebration of Science and Humanity. *Darwin's 200th Birthday will occur on February 12, 2009; it will also be the 150th Anniversary of the publication of his famous book,* On The Origin of Species. *So, together we can evolve a truly international Celebration to express gratitude for the enormous benefits that scientific knowledge, acquired through human curiosity and ingenuity, has contributed to the advancement of humanity. The objective of Darwin Day Celebration is to encourage existing institutions worldwide, such as municipalities, public and private schools, colleges and universities, libraries, museums, churches, private organizations and individuals to celebrate Science and Humanity every year, on, or near, February 12, Darwin's birthday!* [1]

This sheds light on the relentless fight on the part of these devotees for Darwin and his theory of evolution. Science, itself, does not call people to action, but belief in Darwin does. They are mounting a humanistic religion.

Evolutionists will not tolerate a challenge to something that has become such an integral part of their lives. That is why they fight against anything that challenges evolution. The Darwin Day website goes on to say:

> *Celebrations are an important part of every culture. They provide a tradition and a common bond to be shared among those who make up their culture, permitting them to experience a meaningful connection to one another and to the principles to which they subscribe. Unfortunately, most celebrations are based on ancient traditions that are relevant to only a specific country or culture, and they have often been, and continue to be, the source of serious conflicts.*
>
> *At this juncture in history, the world has become so small and interdependent that we need a Global Celebration to promote a common bond among all people. The Darwin Day Celebration was founded on the premise that science, like music, is an international language that speaks to all people in very similar ways. While music is both intellectual and entertaining, science is our most reliable knowledge system, and it has been and continues to be acquired through human curiosity and ingenuity. Moreover, evolution via genetic variation and natural selection, introduced by Darwin, has become the central organizing principle in biology.[2]*

Evolutionist's want a "Global Celebration" that will be a "tradition and a common bond"! Which part of this is science? They have created a religion that worships human ingenuity. They *are* calling it a Global Celebration and they *will* try to eliminate other traditions that become what they call "sources of serious conflicts," i.e. Christmas, Easter, Good Friday, Sunday worship services, and praying before meals.

This is not science. Only a worldview or religion espouses unity, celebration, or the intolerance of other beliefs. No other "scientific" theory enjoys this level of popularity and homage. The theory of evolution is decidedly different and separates itself as a different phenomenon from other scientific theories.

We are now witnessing great signs of the times. Nearly every core sociological principle of evolution leads to great sin. Therefore, is it really a surprise that the full acceptance of this ideology comes with a prerequisite to reject the idea of sin altogether as we learned in the Sub-Law of the Doctrine of Sin? A discerning mind will see that this alternate creation story carries the fingerprints of evil—it does not matter what gift-wrapping it comes in: science, philosophy, or Global Celebration.

Besides their efforts to create a Global Celebration centered on Darwin, evolutionists employ another tactic commonly used by religion to win coverts: they spend unbelievable amounts of time attacking their competition.

Evolutionists continue to hate Jesus without cause, just like the Pharisees did. They have written numerous books on the subject including: *The God Delusion* by Richard Dawkins, *God: The Failed Hypothesis* by Victor J. Stenger, *Evolving God* by Barbara J. King, *God is not Great* by Christopher Hitchens, and many others. For people who say God does not exist, they seem to spend an awful lot of time arguing with Him!

It almost seems like their attacks on Christ outpace their desire to support evolution. It is possible they are more obsessed with God than many people who call themselves Christians. This is another sign of a false religion, for they are jockeying for power against Christ, with philosophies and empty deceit. (See Colossians 2:8.)

These attacks are plentiful. Many of them take the form of books like the above-mentioned *The God Delusion*, *God—The Failed Hypothesis*, *God Is Not Great*, and so on.

However, their attacks on Jesus and His followers put evolutionists in a fascinating predicament. Jesus was very clear when He said that being His follower would bring persecution. In a very real way, evolutionists, who deny the content of the Bible and ridicule Christians for being archaic and stupid, are fulfilling the Bible's very own words:

And you will be hated by all for my name's sake.
— MATTHEW 10:22

No matter how much they try, evolutionists will never get away from Jesus, nor will they stop fulfilling His words. That is the great irony of their cause. If they want proof that Jesus Christ is real, they should look at their own lives and actions as they mock, ridicule, and persecute Christians. If they can be honest with themselves, they will see how their very actions mirror the words and prophesies in the Bible.

The flip side of this argument is even more interesting. If evolutionists want proof that evolution is true, they cannot look at their own lives and actions; they have to look at hypothetical situations they imagine occurred billions of years ago.

To be thorough in this discussion, we need to look at evolution through spiritual eyes. It will take spiritual discernment to understand why the philosophy of evolution is here and ascertain what we can do against its reckless hatred.

Such a discussion certainly must have Satan in it. As a Church, we have given him way too much pulpit time and credit for our own shortcomings. Too often we overlook our own sinful nature and our capability to do awful things, and simply blame it on the enemy. But evolution is different. We must understand a few characteristics of Satan to begin to understand evolution's end game.

Satan is interested in large-scale, strategic assaults upon God and His people. He is too limited to be involved in the rather small, personal struggles we each deal with—his plans are on a much broader scale. As humans, we don't need the prince of darkness to woo us into disciplinary issues. He is onto bigger things. He is not trying to rob a convenience store; he is trying to hijack the Federal Reserve, so to speak.

THE END GAME OF EVOLUTION

Satan is not God so he does not think about each of us individually. Many times, we assign attributes to him that do not belong to him. When we think of the archangel Michael, we don't think of a being whose presence is everywhere and has the ability to influence each of us personally for good. God assigned those attributes solely to Himself.

In a sense, however, we assign those attributes to Satan. We often think he is everywhere, that he personally tempts each one of us, and that he is the very powerful, almost equal adversary of God. Unlike God, his power is limited because he is neither omnipresent nor omniscient. God, on the other hand, has the unlimited power to be involved in every minor detail of our lives.

Satan is actually the same kind of creation as the Archangel Michael, only fallen. Therefore, Satan is not even close to being capable of tempting each one of us on the planet and being involved in each of our lives on a personal level. Instead he has to focus his time on things he *can* do.

We find an illustration of the nature of Satan in 1 Peter 5:8:

> *Be sober-minded; be watchful. Your adversary the devil prowls around like a roaring lion, seeking someone to devour.*

This illustration depicts a singular being that occupies a certain amount of space, but he is on the move, like a lion hunting for prey. The reason he is on the move is that he can't be everywhere at once.

This verse is also enlightening because the lion's best tactic for hunting is to cause fear in its potential prey. Once a lion causes panic in a herd of zebra, the herd runs and the weakest members of the herd fall behind and make easy prey.

What is revealing about evolution is that people who ascribe to it often panic about the stability of the world. We can see this plainly in some of the environmentalist campaigns and expressed thoughts. In an extreme case, author Lamont Cole has been quoted as saying that "To feed a starving child is to exacerbate the world population problem."[3]

THE EXTINCTION OF EVOLUTION

This kind of thinking just does not make any sense to God-fearing people. Yet we can begin to understand its foundation when we institute evolutionary thought. We remember that Eric Pianka thought that the world would be better off with 80 to 90 percent of us gone.

This panic (instigated by population fears) derives from the evolutionary thought that we are accidents of nature. Therefore, our future could accidentally be eliminated in another, unexplained, unplanned way. In this evolutionary-induced state of mind, all sensibilities are vacated and fear of the unknown becomes the governor of decision-making.

These people who are stapled to the millstone of evolution and not grounded in the Word of the Lord, will be easy prey for Satan to pick off. However, unlike a lion that kills and devours immediately, the prey of Satan become the walking dead. They continue to move around completely lifeless in their transgressions, and with mouths that are like open graves, they shout that there is no God.

> *Their throat is an open grave; they use their tongues to deceive. The venom of asps is under their lips.*
> — ROMANS 3:13

Another example of the limited abilities of Satan and his demons is found in the Book of Acts in which seven brothers tried to cast out a demon from a man. The brothers invoked the names of Jesus and Paul (even though they truly did not believe), and in Acts 19:15 the demon responded in a most enlightening phrase, "Jesus I know, and Paul I recognize, but who are you?"

Here we see men trying to cast out a demon, and the demon did not even know who they were. But the demon had heard about Paul and he personally knew Jesus. Even though that demon was not Satan, we can relate his limited characteristics to Satan, for they are similar beings.

So, no, Satan does not even know who many of us are. He neither has the ability, nor the desire to get involved in our

THE END GAME OF EVOLUTION

individual lives. However, in saying that, he is obviously far from being inept. He is the master of the domino effect.

Satan pushes over strategic dominos that lead to societal collapse. His plans are immense and scale generations of humans, not just a few years. He is trying to influence mass killing, cultural abominations, hatred of God, and widespread moods and beliefs. Evolution certainly fits within that playbook.

Another page of that playbook can be seen in the concerted attack on the sanctity of the basic family unit. That shows another example of a wide-scale affront on a Christian institution.

Due to Satan's inherent limitations, he must elicit the help of mankind to accomplish his goals: to inflict as much pain as possible on everyone he can. But there is no victory against God, so any amount of pain inflicted by Satan will ultimately be returned to him many fold in his ultimate judgment. In essence, he is only hurting himself in his hateful efforts; all who fight against God will travel the same paths.

Despite that inevitable failure, insanity is the brother of hatred. Therefore, in his insanity, Satan presses forward in a battle he has already lost. We see the same attributes of insanity and hatred in evolution. One could say unbridled hatred and insanity are the signatures of Satan, and Charles Darwin was simply the perfect parchment upon which Satan could script his masterpiece.

From what we have learned about Darwin, we know he was a bigot. He thought indigenous people were miserable, wretched savages. He thought some of them were subhuman, and others, not even human at all. That is staggering hatred.

Darwin said women have lower mental capabilities than men and are altogether inferior. He thought their powers of intuition and imitation likened them to monkeys and therefore linked them to a lower form of life.

Darwin was soaked in self-pride to the irrational, ludicrous, and delusional end that he believed evolutionary forces made the members of his privileged social class all better looking than those of lower financial means. If nothing else pointed to his insanity, this would be enough to indict Charles Darwin of being mentally unstable.

THE EXTINCTION OF EVOLUTION

Real pain filled Darwin's life. He suffered the devastating loss of his ten-year-old daughter, and at that point, his faith in God was stripped away. He was filled with rage at God for allowing his daughter to die. I think it is fair to say he hated God.

Satan looked at Darwin's character and said, "This is someone I can use." Satan was interested in using him to begin a worldwide assault on God, and he used evolution to facilitate such an assault.

Satan needed evolution because the historical figure of Jesus is well-documented and the person of Jesus is well-respected in today's world—even by people who don't believe Him to be the Savior. He advocated loving your neighbor and loving your enemy; even when those who hated Him were crucifying Him, He prayed for their forgiveness. Jesus Christ embodied these morals so well, that even atheists today, like Dawkins, invoke the Golden Rule (which is Matthew 7:12) as a way that we should all want to live (obviously we see some consistency issues when atheists quote Jesus).

Therefore, Satan needed something more deceptive and subtle to eliminate Jesus. That is where evolution comes in. Evolution does not directly attack Jesus, it attacks Adam, which is a sinisterly clever maneuver; something beyond the mere human ability of Charles Darwin.

The Sub-Law of the Doctrine of Sin is one of the most important chapters of this book. It details the sophistication of Satan's attack. Satan thought his best shot at doing damage to Christ was by attacking Adam and Eve, because for many people there is an untouchable vagueness and unfamiliarity surrounding these first human beings. Satan is staking his claim on the hope that maybe he can simply make us believe Adam and Eve never really existed and thereby effectively shut down the gospel.

Many of us can understand Peter and his denials. We can see the strengths and failures of King David. We can admire Stephen as he was martyred. But Adam walked around naked and had a wife who was made out of one of his ribs. Relating to that can be difficult.

That is the evil genius behind evolution. As we learned through the Doctrine of Sin, evolution teaches that mankind is descended from some ape-like parent rather than a historical Adam and Eve. This is where evolution became too complex for Charles Darwin to fully compose on his own.

Darwin was already predisposed to removing the idea of God from everything, along with the predisposition to believe he was more advanced than other races of men. This bias made the theory of evolution very appealing to him, because it fed his pride—it made him the most advanced being on the planet.

I do not think, however, that Darwin realized that if people believed man descended from an ape-like creature instead of Adam, then Jesus would be made completely obsolete. I don't think he had the spiritual intellect to understand that without Adam, there can be no original sin. And if there is no sin, there is no need for salvation from that sin, and therefore no need for Jesus.

No man could have orchestrated such a massive sociological movement to remove Jesus from every discussion and do it in such a covert, sophisticated way that he never even mentions Christ's name. Furthermore, evolution does not just remove Jesus, it removes all love and morality, and it does it oh, so quietly.

Evolution is the most thoughtfully planned attack on Christianity the world has ever seen. This was not the brainchild of a man; Darwin was merely the pawn in Satan's game. His science was the shiny mirror carrying the cocaine.

The fact that this plan has entrapped many must delight Satan. These evolutionists who have been made to be in fellowship with God have concluded, through their collected human insight, that they are nothing but worthless, purposeless, accidental, and unavoidably mortal animals.

Evolutionists have adopted a belief that begins and ends with the impending question of when their inevitable death will arrive. In their best possible outcome, their future is a decaying corpse and their life force will be erased from existence along with everyone that they ever loved, knew, or cared about.

Each one of them undoubtedly fears death, because they will always, in the depths of their minds, worry about being wrong about God. There is no upside to their creation story. In fact, it is not really a creation story; it is a story of meaningless life leading to a certain death. Yet, somehow they embrace it, and they boast about the exalted intellect that led them to these conclusions.

Christians however, have tasted the living water. We know the Holy Spirit and the joy and cleanliness in a regenerated heart. We know there is good, and that good is Jesus Christ. It is not speculation, it is not blind belief, nor is it just because the Bible tells us so. We believe because of the supernatural intervention of God in our lives that we cannot deny. From coast to coast, country to country, culture to culture, millions testify to this salvation, and it is real, tangible, and life-changing.

Through the ideology of evolution, we can see how separating oneself completely from God not only renders life hopeless and void of any explanation of love, but it also ends in the final violent demise of the individual in every situation.

Yet to cover that gaping wound with the equivalent of a band-aid, evolutionists have appropriated for themselves empty days of religious-like celebration and claim pseudo-divine concepts for themselves. One website promoting atheism as a religion actually states they are "winning souls for Darwin."[3] Another article, entitled "Darwin's Natural Heir" described evolutionist E.O. Wilson as "ascending to the right hand of Charles Darwin."[4] They do all these things to somehow bring value and purpose to their own lives, even in the wake of admitting their own ideology claims there is no value to any life.

Even still, evolutionists are actively trying to win souls for Darwin and expect Christians to drink from this self-annihilation ideology. But they have no idea that what they are offering are jugs of sawdust when we already drink from the spring of eternal life.

Evolution is becoming more militant every day. The term godless, as shocking as this will sound, is fast becoming a buzzword. There is even a book titled *Godless in America*. Godlessness is literally being flaunted.

This mentality of godlessness is very important for us to understand. If atheists are truly atheists then that would mean they

THE END GAME OF EVOLUTION

do not believe in the existence of God. Therefore, they should already feel like they have a godless America.

Now, I understand that consistency is not their strong suit. I actually wonder if atheists even know what they are saying. Because to believe them at their word as they are campaigning for a godless America, they become one of the most vocal, bigoted groups in modern American history.

From an atheistic perspective, wanting a godless society can mean only one thing: atheists don't want people who believe in God to live near them or in any way intersect their lives. They cannot kick what they think is a non-existent God out of America; they can only hope to drive people out of America who believe in God. If they really do not think God exists, He cannot be their enemy—those who have faith in God are their enemies.

Atheists are exhibiting the same human sin born from ignorance that leads to racism; they have just modified it. They have isolated a group of people to attack because they have a characteristic that is different from them. Only this time, it is not skin color—it is a belief in God. It is our belief in Jesus Christ as our Lord and Savior that has made us a target of atheistic animosity and they do not want us here. The atheists are doing exactly what that Bible prophesied they would do:

> *If the world hates you, know that it has hated me before it hated you.*
> — JOHN 15:18

Atheists have decided that belief in evolution, and the blind force that luckily created us, is reasonable, but faith in a living God is appalling. They are adamant that America must become a godless nation. From their perspective, the only way to become godless is to eliminate, in one way or another, all who believe in God because we are the only face of God that they see.

They probably believe that the easiest way to make America godless is to get everyone to denounce God. However, will they stop pursuing a godless America when they realize we will not denounce God because we know He exists? The answer is no.

This leads us to an amazing point. Our very lives, when they are layered in the fruits of the Spirit, completely disprove evolution. And it is that living proof of Christ in our lives that offends the evolutionists the most. The more Christ-like we become, the clearer it is that survival of the fittest could not have created us, and it is that face of Christ that is hated so much by the atheists.

This friction is going to heat up to volcanic levels—levels that were foretold in the Bible! Christians will become like an unwanted pregnancy. At that point, what will the atheists' ideology of evolution through natural selection allow them to do? It allows them to do anything they can get away with—from abortion to euthanasia—and every kind of persecution in-between.

The evolutionist ideology has already had a test run with Adolf Hitler, so we know people have the ability to rally behind a charismatic leader who embraces this doctrine. We are not talking about a hypothetical situation; the train tracks have already been laid and there has been an inaugural run.

There is an even deeper problem, than those we have already discussed, for the evolutionary atheists who are crying out for and wishing to be in a godless society. When an atheist calls for a godless society, what they are really asking for reveals the most terrifying state of human existence we can imagine.

There is only one place throughout time and space that is fully, and completely, away from the presence of God, and that is hell. (See 2 Thessalonians 1:9.) The atheist who is completely consumed with evolutionary thought, and is driven by evolutionary ideology, has a soul so broken that they actually broadcast that they want to be in hell.

In his skillful exposition of the Psalms, noted theologian Charles Spurgeon provides a spotlight on the exact issue we are dealing with. Spurgeon wrote:

> It is the psalmist's desire to teach us the way to blessedness, and to warn us of the sure destruction of sinners.[5]

This intention is evident in the very first verse. He begins with quoting Psalm 1:1, KJV:

> *Blessed is the man that walketh not in the counsel of the ungodly, nor standeth in the way of sinners, nor sitteth in the seat of the scornful.*

With gifted wisdom, Spurgeon helps us understand the power in this verse. He writes:

> *Mark the gradation in the first verse. He* walketh *not in the counsel of the ungodly, nor* standeth *in the way of sinners. Nor* sitteth *in the seat of scornful.*
>
> *When men are living in sin they go from bad to worse. At first they merely walk in the counsel of the careless and ungodly, who forget God—the evil is rather practical than habitual—but after that, they become habituated to evil, and they stand in the way of open sinners who willfully violate God's commandments; and if let alone, they go one step further, and become themselves pestilent teachers and tempters of others, and thus they sit in the seat of the scornful. They have taken their degree in vice, and as true Doctors of Damnation they are installed.*[6]

Spurgeon gives us a powerful biblical truth about the downward spiral of sin: sin leads to more sin. Sin cannot lead to righteousness and is never satisfied with the status quo. Once a new threshold of sin takes hold the sinful nature lusts for more until the full destruction of the sinner is accomplished. Only the redemptive power of saving grace can turn the sinner's downward spiral around and bring righteousness and truth to that heart.

I understand some readers will be offended that I group those who believe in evolution through natural selection with atheists. I do so for two reasons. First, if you maintain a belief in both Jesus Christ and in the doctrine of evolution, you understand neither of the doctrines. Hopefully I have shown that the two are completely incompatible.

Secondly, if you partially entertain the first few notions of evolution, you are beginning the rejection of God's own account of His creation. You may have not paid yet, but you have invited the prostitute into your house. That is the first step of walking in the counsel of the wicked.

That first step of evolution is the very definition of a slippery slope. The belief that mankind descended from an ape-like creature irrefutably puts one in the same camp as the atheists. It is that false foundation that makes things go from bad to worse. The atheist simply represents the natural ending conclusion to that initial evolutionary thought.

This gradation of evolution, ending in the militant atheist, becomes a stunning example of Psalm 1:1. In *The Extinction of Evolution*, we have explored this evolutionary descent and watched it go deeper and deeper into sin. We are now going to condense it, and view the tumble in one collective burst. Doing so will provide a clear picture of the pure evil Satan is perpetrating. We will show what he is able to accomplish through subtle deceit, half-truths, and the numbing effect of time.

Evolution contradicts God's creation account of Adam and Eve. In doing so, evolution leads to the elimination and nullification of the entire idea of sin that came at the hands of Adam.

Evolution reduces the Savior from sin, Jesus Christ, to some man who happened to be executed. Without Adam, and a spiritually significant Jesus, the Bible ceases to mean anything beyond mythology.

Without the Bible, there ceases to be a reliable source that proves there is a God, so God must not exist. Therefore, man ceases to be a creation and becomes an accidental organism with an aimless direction and no inherent value.

Evolution leads to a society where love, kindness, and charity should naturally not exist, because it is a conundrum that cannot be explained through natural selection. Evolution makes compassion a waste of time, adoption a confusing event, and abortion a natural solution.

Evolution leads to people crying out for a godless society. The only place, in all of time and space that is completely godless is hell. Therefore, evolution leads people to literally ask for hell. It's no wonder Satan is called "the deceiver of the whole world." (See Revelation 12:9.) This slide of massive proportions started with the quiet lie that Adam never really existed.

If redemptive grace does not come to the people who follow this ideology all the way to the end, not only will God grant them their desire for a life separate from Him, He will do so for all eternity.

That is where the extinction of evolution occurs and where the belief in evolution will forever cease. It will only happen when the evolutionists are given everything they have so passionately asked for—a place completely void of God and His people.

Charles Darwin was one of the first to walk that long road and follow his theory to the very end. When he bottomed out on his awful maturation of sin, he called the saving message of Jesus Christ "a damnable doctrine."

No, Mr. Darwin, Christianity is not a damnable doctrine as you have claimed. Jesus Christ came to save the world and evolution certainly came to kill it. For in evolution, there is no hope of joy, no promise of life, and no love of a Creator. There is only a guarantee of death.

In contrast, while evolution ends in ultimate death, Christianity offers ultimate life. Those who call upon the name of Jesus Christ, the Savior of all who love Him, can live in the presence of God for eternity.

ENDNOTES

1   http://www.Darwinday.org

2   http://www.darwinday.org/darwin_symbol.html

3   http://www.freerepublic.com/focus/f-news/1145419/posts

4   www.churchofreality.org

5   Douglas, Ed, "Darwin's Natural Heir," Guardian Unlimited, February 17, 2001.

6   Spurgeon, Charles, *The Treasury of David*, Hendrickson Publishers, pgs.1-2.

CHAPTER SIXTEEN

SALVATION IN JESUS CHRIST

Everyone has sinned against God. (See Romans 3:23) We see the results of our sin all around us: war, greed, lust, and all levels of strife and stress. (See Romans 3:15-17.)

But what exactly is sin?

Sin is not solely an act of cheating, lying, stealing, or using God's name in vain. Sin is also evil or impure thoughts and motives. Sin describes the natural disposition of the human heart. A person does not become a sinner because they commit acts of sin. We commit acts of sin because we are naturally born sinners.

We are born in a state of depravity because of the original sin at the hands of Adam. (See Genesis 3.) That sin of Adam was one of pride that manifested itself through disobedience. Adam chose his own way and disobeyed the law of God. That is the original sin that has cursed humanity ever since.

Subsequently, we are all descendants of Adam, and we inherited his rebellious sinful heart. This sin is visibly seen in our immoral ways and it is a crime that humanity is guilty of before the perfect and righteous God.

The judgment for the crime of sin is death. (See Romans 6:23.) Sin is a capital offense and God sits as the all-knowing, non-erring judge. His judgment of sin is an eternal death, which is hell.

We cannot make amends for our crime of sin. We are not capable of fixing what is broken. (See Romans 3:10-20.) We stand helpless and completely guilty before an almighty and just judge who knows every evil desire that we have ever had.

However, in an unprecedented move, God took off the judge's robe, and became a substitute for the guilty sinner. God subjected Himself to the shame, disgrace, and pain of standing trial for all of us, and accepted the punishment of death for all of humanity. He did this through Jesus Christ.

Jesus Christ was and is God, manifested as a person, to be the sacrifice and to pay the sinner's penalty of eternal death.

To pay the debt, Jesus Christ had to be born from a virgin; for the curse of sin is passed through the seed of Adam, the seed of the man. The virgin birth was God's supernatural intervention to have a child born not naturally a sinner, and to demonstrate that the child was indeed Immanuel, meaning God is with us. (See Matthew 1:23.) Then Jesus Christ lived a life free of sin, and He himself was worthy of eternal life based on His merit. Death had no hold or legal claim to the life of Jesus Christ because He never sinned.

Yet, because God so loved the world He sent Jesus Christ to be the sacrifice for the world. (See John 3:16.) The sins of humanity, past, present and future, were levied against the perfect Christ. (See 2 Corinthians 5:21.) The penalty of our sin was paid by the blood and death of Jesus.

But because Jesus himself was blameless, death could not hold him. Therefore, Jesus Christ was resurrected from the death he suffered as the result of His crucifixion. His Resurrection is the most awesome event in the history of mankind. Christ's Resurrection defeated the sin that came into the world at the hands of Adam. (See Romans 5:18-21.)

It took every aspect of the virgin birth, the sinless life, the death on the Cross, and the glorious Resurrection of Jesus Christ to make salvation from sin possible. Therefore, Jesus Christ is the only one who can grant salvation from sin because He is the only one who is capable of paying that debt. The road to salvation is narrow. Jesus Christ is the only way, the only truth, and the only life. (See John 14:6.)

It is not an empty incantation of words that obtains salvation. It is not a ritual, and it has nothing to do with culture. Church attendance cannot offer it. Works cannot acquire it. Salvation in Jesus Christ is a gift from God. It is the quickening of a heart dead to sin and regenerating it with the life of Christ.

This gift brings with it a longing of the soul to know Jesus. A repentance from acts of sin and of sin nature and an understanding of its punishment. It draws you to an honest confession that the Resurrection of Jesus Christ is the salvation for your soul. It brings a commitment and a devotion of your life, your soul, and your eternity to your King, Lord, God and Savior, Jesus Christ.

Indeed, it brings a promise and a guarantee.

> *If you confess with your mouth that Jesus is Lord and believe in your heart that God raised him from the dead, you will be saved.*
> — ROMANS 10:9

And if you do so with a truly repentant heart, you can be assured of your eternal salvation and a life that never ends in the Kingdom of Heaven. For God promises His children they will never be separated from Him.

> *For I am sure that neither death nor life, nor angels nor rulers, nor things to come, nor powers, nor height nor depth, nor anything else in all creation, will be able to separate us from the love of God in Christ Jesus our Lord.*
> — ROMANS 8:38-39

A declaration is now being asked of you. Will you confess Jesus Christ as your Lord and Savior, or do you choose to reject Him?